(27.95)

No Nails, No Lumber

No Nails, No Lumber

The Bubble Houses of Wallace Neff

JEFFREY HEAD

PRINCETON ARCHITECTURAL PRESS · NEW YORK

Build thee more stately mansions, O my soul,
As the swift seasons roll!
Leave thy low-vaulted past!
Let each new temple, nobler than the last,
Shut thee from heaven with a dome more vast,
Till thou at last art free,
Leaving thine outgrown shell by life's unresting sea!
—

from "Chambered Nautilus," by Oliver Wendell Holmes, 1858, found among Wallace Neff's personal notes and a primary source of inspiration for his development of bubble house architecture

Sea shells have influenced the designs to some extent as they are one of the most nearly permanent and durable structures for living things. Developed by nature throughout the centuries to withstand tremendous pressures and hazards of the sea, these shapes offer mankind protection from natural destructive forces as well as new man-made forces.
—

Wallace Neff, undated

previous, page 2
A double-bubble house, 1942

previous, page 4
Wallace Neff, in front of the first completed Airform house in Falls Church, Virginia

Published by
Princeton Architectural Press
37 East 7th Street
New York, NY 10003

For a free catalog of books, call 1-800-722-6657
Visit our website at www.papress.com

© 2011 Princeton Architectural Press
All rights reserved
Printed and bound in China
14 13 12 11   4 3 2 1   First edition

No part of this book may be used or reproduced in any manner without written permission from the publisher, except in the context of reviews.

Every reasonable attempt has been made to identify owners of copyright. Errors or omissions will be corrected in subsequent editions.

Editor: Dan Simon
Designer: Paul Wagner

Special thanks to: Bree Anne Apperley, Sara Bader, Nicola Bednarek Brower, Janet Behning, Fannie Bushin, Carina Cha, Tom Cho, Penny (Yuen Pik) Chu, Russell Fernandez, Jan Haux, Felipe Hoyos, Linda Lee, Jennifer Lippert, John Myers, Katharine Myers, Margaret Rogalski, Andrew Stepanian, Joseph Weston, and Deb Wood of Princeton Architectural Press
—Kevin C. Lippert, publisher

Frontispiece: Excerpt from "The Chambered Nautilus," is reprinted from *The Complete Poetical Works of Oliver Wendell Holmes* by Oliver Wendell Holmes (Boston: Houghton, Mifflin), 1895.

Library of Congress
Cataloging-in-Publication Data

Head, Jeffrey.
No nails, no lumber : the bubble houses of Wallace Neff / by Jeffrey Head. — 1st ed.
     p. cm.
Includes bibliographical references.
ISBN 978-1-61689-024-7 (alk. paper)
1. Neff, Wallace, 1895–1982—Criticism and interpretation. 2. Concrete houses. I. Title. II. Title: Bubble houses of Wallace Neff.
NA737.N37H43 2011
720.92—dc23
                                                            2011022740

CONTENTS

8	Foreword, Steve Roden
13	Introduction

The Bubble Houses (USA)

34	Falls Church, Virginia
42	Litchfield Park, Arizona
52	Loyola University, California
54	Pacific Linen Supply Co., California
60	The Andrew Neff House, California
68	South Pasadena, California
72	Hobe Sound, Florida

Airforms Around the World

86	Latin America
104	Europe
114	Africa
122	Asia

APPENDICES

127	A – Interview: Former Bubble House Residents
137	B – Patents of Wallace Neff
146	C – Selected Unbuilt Airform Projects, 1944–1958
158	Afterword
160	Acknowledgments
162	Notes
165	Selected Bibliography
168	Image Credits

FOREWORD
Steve Roden

THE OVAL HAS LANDED
One turns onto a small residential street and is confronted with a house that looks nothing like a house. The fact that it has a front door and a few visible windows only adds to its incongruous presence amidst the traditional 1930s homes that comprise most of the neighborhood. Looking at the house, which resembles a smooth mound of earth, it feels as if some ancient space station has suddenly fallen from the sky; and upon landing, it has mysteriously embedded itself into the wrong context, its presence so strange it seems to have traveled through both space and time. Sixty plus years after it arrived, no one can deny that this dome, with its oval footprint, still retains a tremendously uncanny presence: it looks like it could be a home...but is it?

HANDFORMS, MANFORMS, AND AIRFORMS
Standing in the central living room of Wallace Neff's last remaining bubble house, one can see that the inside of the dome is hardly perfect. In bright light one can see its inconsistent surface, which

has no resemblance to anything made by a machine. The surface of the dome, as seen from the inside, looks like it were modeled by hand, although in truth it was formed by air. This interior surface is decidedly human; in certain daylight hours it feels less like a cover, and more like an opening. While the house feels bogged down on the outside, it feels soft as an atmosphere on the inside.

IN PRAISE OF SHADOWS

On cloudy days its interior remains a bit dark; I assume that the lack of windows has something to do with the structural necessities of the shell. Nevertheless, light penetrates the windows and doors in wonderfully mysterious ways. The curved interior and freestanding walls allow refracted beams of light to enter the house, projecting abstract shadows that move across a variety of interior surfaces. As sunlight bounces images off of passing cars and sprinkler sprays, the light and its accompanying shadows resemble some of Hans Richter's abstract films of the 1920s. Like insects, these images appear randomly at all times of the day, and generally in a variety of locations. At night the films also appear, projected via headlights, creating colored, flickering presences, as if Neff had hidden a series of Victorian magic lanterns in the garden.

BREATHFORMS

After living in the bubble house for a few years, I began to think about how the interior space was truly formed of air, how Neff's genius had allowed air to become a sculptor. As much as I had been fascinated with the humble qualities of the dome's inner surface, I began to think about the house in relation to breath. Several years later, I generated a sculpture and sound installation entirely wrought from my own breath, related to the bubble houses that Neff built in Litchfield Park, Arizona, in collaboration with the Goodyear Company. The installation attests to the potential effects a house can have on its inhabitants, which are not only related to aspects of inspired living, but also to inspired making.

INTRODUCTION

Fig. 1: Neff at an Airform construction site

Architect Wallace Neff (1895–1982) is perhaps most well-known for designing large, elegant Spanish Colonial–revival homes, built in Southern California for very wealthy clients. *No Nails, No Lumber: The Bubble Houses of Wallace Neff* is a survey of Neff's lesser-known work: Airform-constructed buildings. Airform construction, a type of pneumatic architecture, uses an inflated balloon to create its form and structure. Neff often referred to these as "bubble houses," and considered them his most significant contribution to the field. He hoped the bubble houses would be his legacy, considered his greatest contribution to architecture. [Fig.1]

Neff's development of the bubble form was largely the result of a hands-on approach that he pursued throughout his life. At a young age he made architectural drawings and eventually taught himself enough about architecture to gain special admission to MIT, but he did not complete his studies. When the United States entered World War I, he chose to return home to Southern California where he worked for the Fulton Shipbuilding Company. This experience introduced him to cement construction,

Fig. 2: A 12-foot-diameter Airform under construction near Burbank, California

knowledge that reinforced his interest in architecture and became central to his work.

After the war, in 1919, Neff began working as a draftsman for George Washington Smith, a Santa Barbara architect who led the movement of Spanish Colonial–revival architecture in Southern California. This was the formal start of Neff's architectural career, one that spanned more than fifty years and included designs for hundreds of buildings. In addition to the estates Neff built for Hollywood's most popular film stars like Judy Garland, Groucho Marx, and Douglas Fairbanks Jr., he also constructed public buildings—libraries and churches.

Neff applied the same intrinsic sense of scale and proportion to both his estate-designed homes and his bubble houses. Although the latter were never intended to be luxury residences, they represent Neff's innovative, industrious nature. The technical details of the bubble houses belie their aesthetically simple design in addition to Neff's engineering skills. His sense of individualism and entrepreneurial interests combined a larger cultural perspective with a desire to service society's basic need for housing.

Bubble houses require no special tooling and their building equipment is portable. Neff would repeat throughout his life that "Airform construction permits the best of modern design for the least money, yet permits building with materials which are plentiful."[1] Beyond the ease of construction and functionality, he also considered the structures to be, "the last word in streamlined beauty and efficiency."[2] The form itself was its shape. Its structural mass was neither hidden nor masked. [Fig. 2]

Many referred to the Airforms as igloos, including residents. Neff preferred to call them bubble houses.[3] To increase appeal, Neff designed several traditional exteriors, including brick or stone veneer over the cement finish, along with a modified roofline made of asphalt and wood shingles.

While the bubble houses may appear to be prefabricated, Neff made it clear that "Airform is a method of construction and is in no sense allied in any way to construction of so-called 'prefabricated' houses." Rather, it was "a revolutionary method providing for a low-cost, labor-saving process of extremely rapid construction of

Fig. 3: A comic Neff clipped from an unidentified newspaper showing Airform development in Los Angeles

permanent houses and buildings."[4] The uncomplicated, singular form of the bubble houses made them easy to build; however, since they were constructed on-site and did not include preassembled components, they could not be considered prefabricated. Robert L. Davison, of the John B. Pierce Foundation, was aware of Neff's bubble houses and may have had them in mind when he wrote humorously, "an engineer is he who can do for one dollar what any fool can do for two dollars."[5] [Fig. 3]

During Neff's lifetime, thousands of bubble houses were constructed internationally, but by 2010, there was only one example of a Neff bubble house left in the United States, located in Pasadena, California. This last remaining bubble house—coincidentally one of the first—was built for Neff's brother Andrew. For a time, the two men lived there together.

Although Neff retired in 1975, he continued to believe in the value of his bubble house architecture and Airform construction, and its uses for the world's unending need for low-cost housing. By this time, bubble houses had been built in more than fifteen countries. Neff hoped for more. For him, the design was still not fully exploited and represented a new approach to construction and building. He did not give up on the bubble house. He continued to spend his personal time and money developing construction variations and new designs.

Neff was unfailing in his search for new materials and compounds to reduce the time for cement to set and provide greater strength. He experimented with different paints, balloon construction, air pressure, size, and shapes. [Fig. 4] He also performed extensive tests and research that included the removal of a wall section from one house that was rigorously tested in a lab. The tests showed the wool in the insulation not only kept the houses perfectly warm and dry, but prevented cracking from expansion and contraction under extreme heat and cold.[6] This further confirmed the uniformly sturdy construction of the houses, regardless of size. [Fig. 5]

Neff's designs remained progressive and were accepted in many areas such as Africa, Mexico, and South America, however, more often issues with local politics, funding, and a lack of

right
Fig. 4: Neff, in his signature white shirt and black tie (center)

bottom
Fig. 5: A two-page advertisement for Goodyear Mechanical Goods that appeared in several national magazines

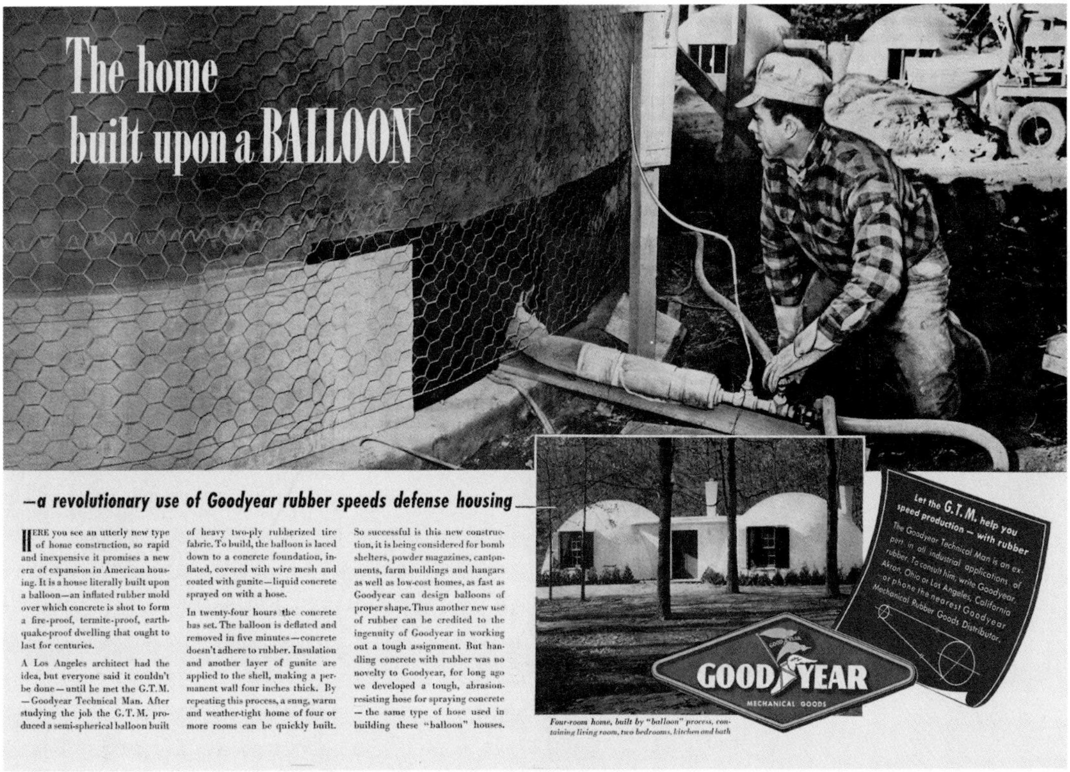

INTRODUCTION 17

diligence from various companies prevented greater implementation of the bubble house. When projects did not move forward, which was a frequent occurrence, Neff returned to building large estates.

Research into Neff's bubble house architecture and other buildings has uncovered their forgotten or previously unknown lineage. Many of the houses were built in areas that remain remote today, others have evolved beyond recognition, and some have been destroyed. There is even a scattered group of houses that were constructed without Neff's permission or awareness. The results were isolated, one-off houses and hybrids. The accounting for all of Neff's Airforms is not complete and requires further study and discovery across the world.

DEVELOPMENT OF AIRFORM

Neff's formal development of the Airform began in 1934, when he made his first drawings of a pneumatic house based on a balloon form. He continued to pursue the concept and wrote a description of his pneumatic form in 1939. For several years Neff pursued alternative construction techniques and reassessed conventional rectilinear design. Prior to his breakthrough design for the bubble house, Neff developed the construction technique for the Rondel House, a low-cost design for cement houses, in 1934. Although his day-to-day client work and finances prevented him from developing the concept further, the process could be best described as a collapsible umbrella sprayed with gunite and then removed from within—a clear predecessor to his Airform design.

Up to this point, engineers in the rubber industry had speculated that it would be impossible to build structures over a rubber balloon.[7] Neff had experimented on several different forms using gypsum; he had studied seashells to understand their convex form and structural integrity; and he had tested balloon seams, reinforcing strips, and their tensile strength. Eventually he developed scale models made in clay that he hand-painted and landscaped, which became the single- and double-bubble houses.

In addition to home construction, Neff believed the utilitarian design had potential uses for aircraft hangers, oil and water

Fig. 6: Neff's set of Airform scale models

storage tanks, hospitals, warehouses, schools, grain bins, and other facilities. In the early 1940s he anticipated uses for Airform construction, particularly during wartime, including ammunition storage, garages, and barns.[8] They could resist weather, explosives, and vermin, and they were cheap and quick to build and easy to maintain.[9] Above all, Neff stressed the simplicity of his Concrete Balloon-Type Buildings, praising their modern sensibilities and seemingly painless manufacture.[10] [Fig. 6]

He called the technique "Airform" and described it as "a new type of construction in which a rubber-coated fabric balloon is blown up and then sprayed with concrete or plastic."[11] [Figs. 7–9] The combination of material, design, and technique was solely Neff's domain. The process was facilitated using gunite, lightweight cement that could be sprayed from a hose. Neff had used gunite in the wood-frame construction of several estate houses he designed, including the Sidney C. Berg Residence (1927), the Arthur Bourne Residence (1927), and the Edward L. Doheny Library at St. John's Roman Catholic Seminary (1940). All three still stand.

The Goodyear Tire and Rubber Company manufactured the first Airform balloons out of industrial-strength neoprene nylon. Although there were no interior load-bearing walls, the strength and weight of the load were evenly distributed across the balloon, which acted as a mold. Costly wood framing was avoided; only minimal scaffolding was required to support workers spraying gunite.[12] A 1941 article in *Life* magazine speculated that even the fifteen pounds of nails used in the construction of each house—an usually small amount—was possibly not necessary.[13]

INTRODUCTION 19

right
Fig. 7: A woman with an ax demonstrating the durability of a built Airform.

bottom left
Fig. 8: A government official attempts an ax test on a built Airform.

bottom right
Fig. 9: A reporter gives an Airform the ax test.

PRECURSORS

While Neff's Airform construction was technically very successful, he was not the first to make the foray into pneumatic architecture, or to explore domes for residential use. Martin Wagner, an architecture professor at Harvard and former town planner from Berlin, Germany, designed his dome-style prefab MW House prototype in 1939. This one-room dome was approximately two hundred square feet and constructed from thirteen curved steel panels. Perhaps the most famous dome architect, Buckminster Fuller, in addition to the geodesic dome, designed the Dymaxion Deployment Unit (DDU) in 1941. DDUs were easily transported, portable dome-shaped structures that were adopted by the U.S. government for military field housing. The 324-square-foot building was divided into three rooms by canvas curtains. [Figs. 10 + 11]

In 1919, British engineer Frederick William Lanchester patented the design for a pneumatic structure for military use in the field as hospitals and other temporary buildings. Lanchester's design used a balloon-form that remained inflated by constant air pressure, allowing people to move in and out of the structure. This perhaps inspired Neff to explore permanent structures based on an inflated form. Meanwhile, construction engineer Fritz Ruppel developed a building system that used concrete reinforced with steel braces to replace traditional wood framing in 1933. This resulted in lower costs and helped to promote the strength of cement against earthquakes and fire.[14] Neff would later adapt the use of steel reinforcement, opting for wire mesh between layers of gunite.

At various times since, architects and designers have revisited the concept of dome housing. Paolo Soleri, working with Mark Mills in 1951, built a dome house in Paradise Valley near Phoenix, Arizona. Their design used two retractable hemispheres that enabled half of the roof to open to the sky. Also in the early 1950s, Los Angeles architect Jeffrey Lindsay applied Fuller's geodesic dome principle to the construction of a house made with ultra-thin, lightweight fiberglass panels reinforced with neoprene—the same fabric Neff used in his Airform balloons. Lindsey's dome,

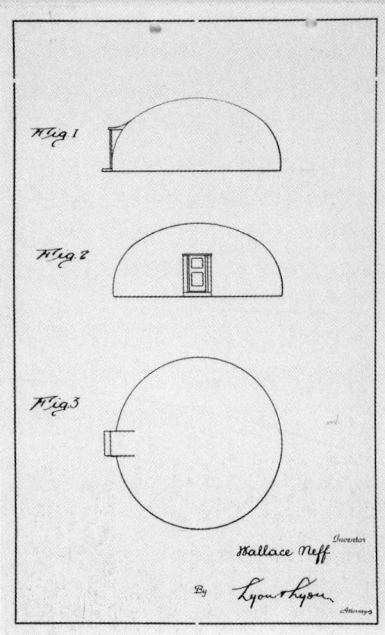 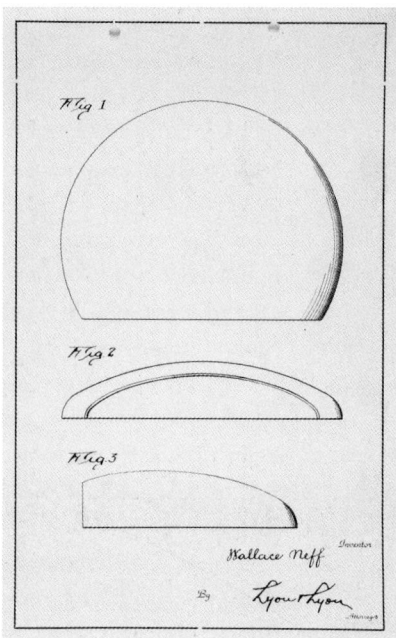

Figs. 10 + 11: Patent drawings

made of fifteen diamond-shaped panels, also included a reflective material for environmental protection.

The countless variations of the dome-home were often considered novelties and regarded as one-of-a-kind structures. Some homes were a demonstration of technical elegance; others offered independence, livability, and an alternative lifestyle. In the United States, no architect, company, or dome-home type was able to successfully transition to widespread cultural use.

CONSTRUCTION PROCESS

The word "construct" comes from the Latin "to pile on," so it is incorrect to say that bubble houses were constructed in the traditional sense. Airform construction was dramatically different from piling brick or stone on top of one another. For one, pressurized air was essential to the construction method. All materials, such as concrete, insulation, or waterproofing were shot into place by a gunite machine using pneumatic pressure, supplanting the slow work of a mason.[15] [**Figs. 12 + 13**]

left

Fig. 12: Photograph of a strength test of an Airform balloon

right

Fig. 13: Detail of an Airform balloon anchored to its foundation

The first stage of construction, like any, was laying the foundation of the structure. Based on the diameter of the Airform balloon, a circular trench was dug at the site and filled with poured concrete to form the footing. Inside the trench, a layer of concrete was poured, filling the ring and forming the foundation and floor. Before the cement set, steel rods were inserted into the foundation and then bent to form hooks.[16] The utility wiring and pipes were then installed.

The balloon was then laid uninflated on the foundation. Grommets mounted along allowed the balloon to be roped to the bent hooks sticking out of the foundation. Once secured, the balloon was inflated through an inlet valve at the bottom. A catenary ring kept the inflated balloon anchored to the foundation. Using

INTRODUCTION

bottom
Fig. 14: A scale model of a single Airform residence

left
Fig. 15: A sketch featuring measurements

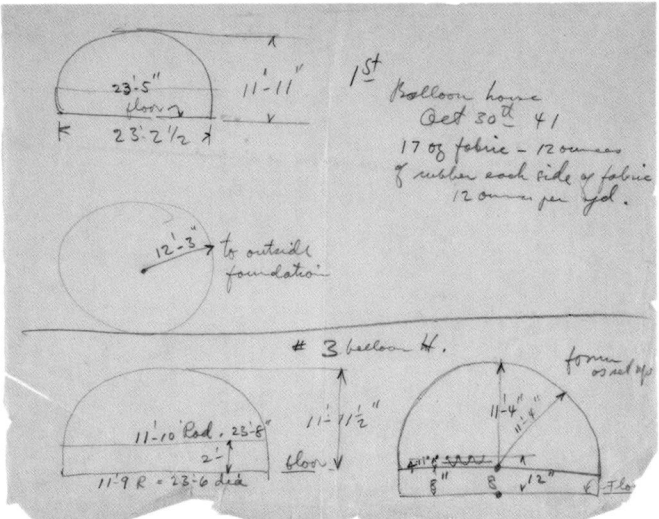

Gunite

Gunite is a mix of dry Portland cement and sand, forced through a hose and forcibly added to water at the nozzle to liquefy the combination on contact. The powerful air pressure required to mix the content must be maintained as the gunite is shot directly from the hose. It is considered to have twice the strength of standard concrete. One of the early modern uses of gunite was in the construction of swimming pools. Initially known as a plastering machine, the gunite process provided structural waterproofing in a variety of uses. It is often referred to as "shotcrete"—shot concrete—and the hose-nozzle as a "cement gun." These became industry terms and were reflected in the names of competing businesses. It became a widely used method of construction throughout the twentieth century, applied to both wood and metal framing. Today it remains an efficient way to produce a dense, compact, and uniform material that is evenly distributed and resistant to moisture. Neff's gunite mix was nearly 3.5 times as strong as ordinary concrete.[20]

Several of Neff's fellow Pasadena architects made early use of the cement gun, including Reginald D. Johnson, with his Harvard Military School for Boys in Los Angeles (1914), and Roland E. Coate with his 1930 addition to his Southern California Automobile Club office building (1923), in the West Adams area of downtown Los Angeles.

approximately one and a half pounds of air pressure through an air compressor, inflation took about five minutes.

At this point, wooden scaffolding was built around the balloon. These temporary wood frames formed the outline for windows, doors, and other openings. The framework had to be sturdy so that the gunite could be shot around it.[17] These openings would also provide access to the deflated balloon once the gunite set. After the wooden framework was erected but before gunite application began, the balloon was coated with a powder to keep the cement from sticking to it. It was then completely covered with a reinforcing wire mesh. After the mesh was laid out, the form was ready for gunite.

Using the cement gun, a layer of gunite was shot directly onto the balloon, from the top down, as the mesh was manually lifted so the gunite reached both sides of the wire. This layer became the ceiling and interior wall of the house. The weight of the top-down technique slightly lowered the ceiling height, which decreased the curvature and made for straighter walls on the side. Thus the domes were not true hemispheres, but more elliptical, with a wider curvature at the top and straighter walls inside.[18] Top-down application helped distribute the gunite evenly.

After the first layer hardened (about eight hours), a one-inch layer of waterproof insulation was applied to the entire form. In colder climates, the insulation was increased to one and a half inches. Neff experimented with different insulation materials that could be sprayed or brushed on, such as pumice and balsam-wool. A Rockwool blanket became the preferred material. When completed, a bubble house would have a thermal value of .16 BTU, an exceedingly low number.

After insulating, the form was covered with another layer of wire mesh and a second layer of gunite was applied. When it hardened, it formed the roof and exterior walls. After twenty-four hours the balloon was deflated and removed from one of the house's openings in the house, immediately ready for reuse. At this point, Neff remarked, "interior fitting and decorating proceeds in the conventional manner." The exterior was painted near the end of the drying stages to absorb more of the paint.[19] [Figs. 14 + 15]

Fig. 16: The *Airform* marketing brochure

THE AIRFORM INTERNATIONAL CONSTRUCTION COMPANY

The completion of several Airform buildings proved the effectual use of the materials and construction method. Based on these achievements Neff's attorneys suggested in 1945 that he turn the Airform Company into a corporation to generate funding and reach a larger market. He called the new firm the Airform International Construction Company (AICC). By 1949 the company had published an extensive, thirty-one page brochure that advertised Airform as "the modern low-cost construction method for concrete buildings," explaining the construction process with schematics, renderings, and photographs of completed work, as well as forthcoming designs.[20] Key projects, such as bubble houses in Virginia and Senegal, storage bins in Arizona, and military housing in Pakistan were included. [Figs. 16–18]

The AICC gradually became more sales-oriented than construction-minded, particulary so in Europe, at the behest of regional director Adolf K. N. Waterval. It was beginning to sink in that without new construction contracts, the AICC would become insolvent.[21] By 1954 the company was preparing to move into a smaller office, having failed to secure projects while conventional buildings remained more aesthetically palatable.[22] The loss of a major commission for the military in 1956 had a permanent impact and the AICC had to fold. It became clear that Waterval was the problem.

An architect based in Alexandria, Virginia, Adolf Waterval designed contemporary brick and stone houses before he became the AICC's regional director in Europe in 1954. Formerly engaged in prefab housing models, he approached Neff a year later about forming a partnership.[23] As director in Europe, Waterval aggressively marketed Airform buildings. He was frequently requesting custom-sized Airforms specifications for potential clients, which required Neff's approval. Neff believed this helped him monitor Waterval's developments. However, as Jose de Lemos, the Airform licensee in Portugal, wrote to Neff, Waterval wanted "barbarous profits at a given moment, not caring for the development of the Airform system, nor having in mind the interests of sub licensees."[24] Waterval would change the terms and conditions with

Figs. 17 + 18: Pages from the *Airform* marketing brochure

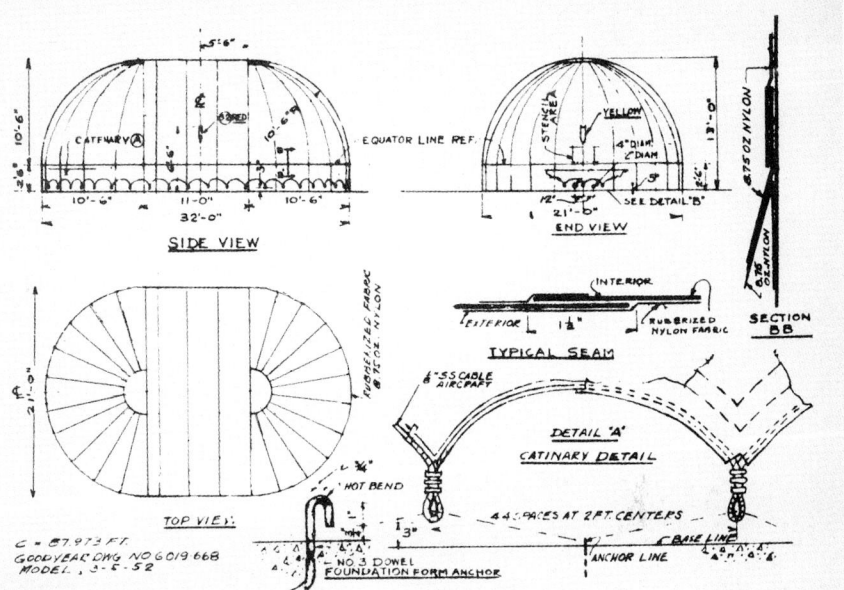

AIRFORM CONSTRUCTION DETAILS
21'-0" x 32'-0" PNEUMATIC FORM

AIRFORM INTERNATIONAL CONSTRUCTION CORPORATION
250 WEST 57 STREET, NEW YORK 19, N.Y.   *telephone* JUdson 6-4223

AC2

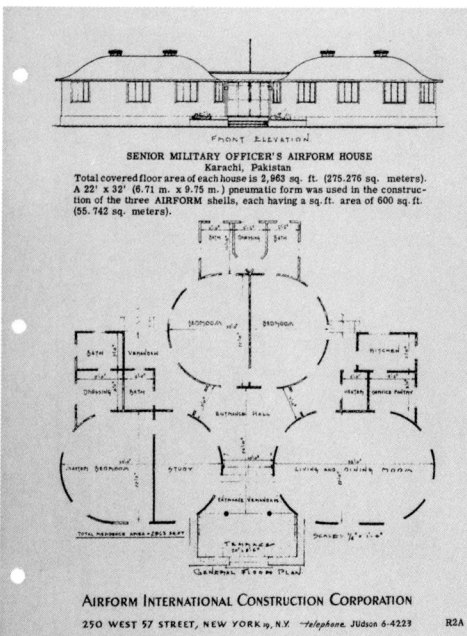
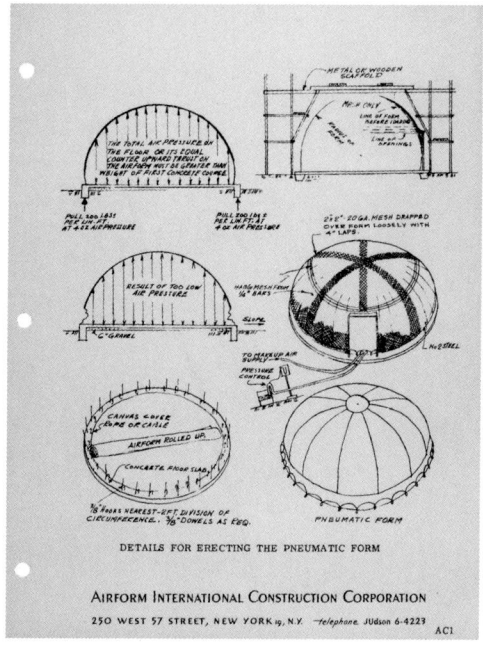
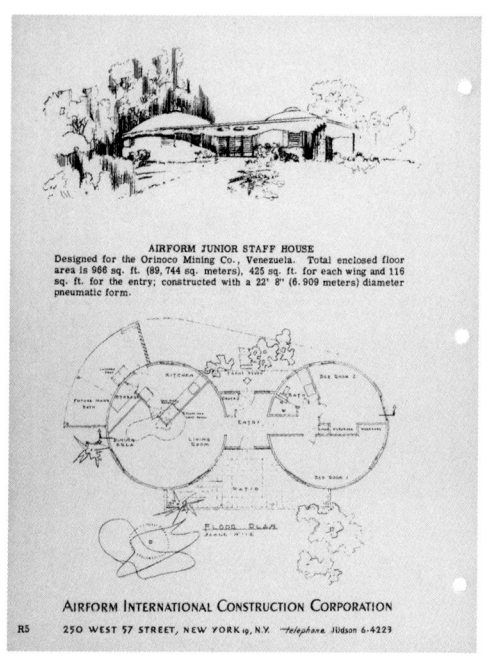
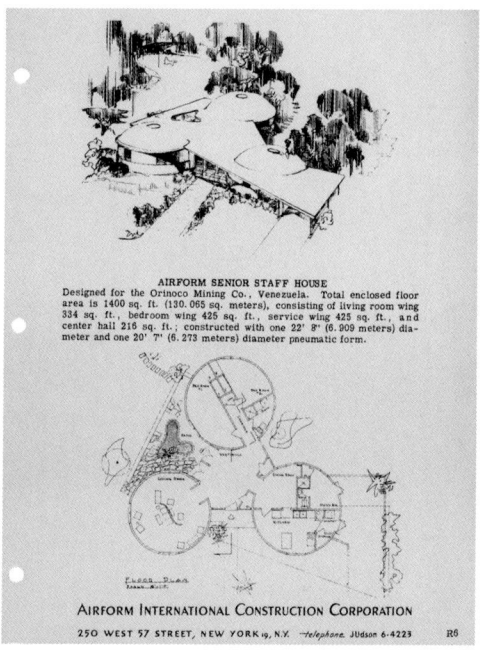

Figs. 19 – 22: Plan pages from The *Airform* marketing brochure

contractors, adding to the debt and growing uncertainty of the solvency of the AICC. He would take advance royalties and write bogus checks. He was likely the cause of several failed projects in Europe during his time with the AICC.[25]

Once the AICC had folded and Waterval was out of the picture, Neff made plans to sell worldwide rights to the Airform construction method. Struggling to find a buyer, he instead decided to start another company, Pneumatic International, Inc. in 1961. Although most of his patents were about to expire, he hoped that Pneumatic International would reactivate interest in Airform construction. Arthur Libby Neff, Neff's youngest son, was president of the new company. He was able to make agreements with licensees in the early 1960s that would pay Neff a royalty fee of twelve and a half cents per square foot of completed Airform space.[26]

Although Airform structures had been built all over the world, eventually the company could no longer compete with other modern architectural developments.[27] It was lost to history in the late 1960s as large-scale housing projects became more institutional and industrialized. [Figs. 19 – 22]

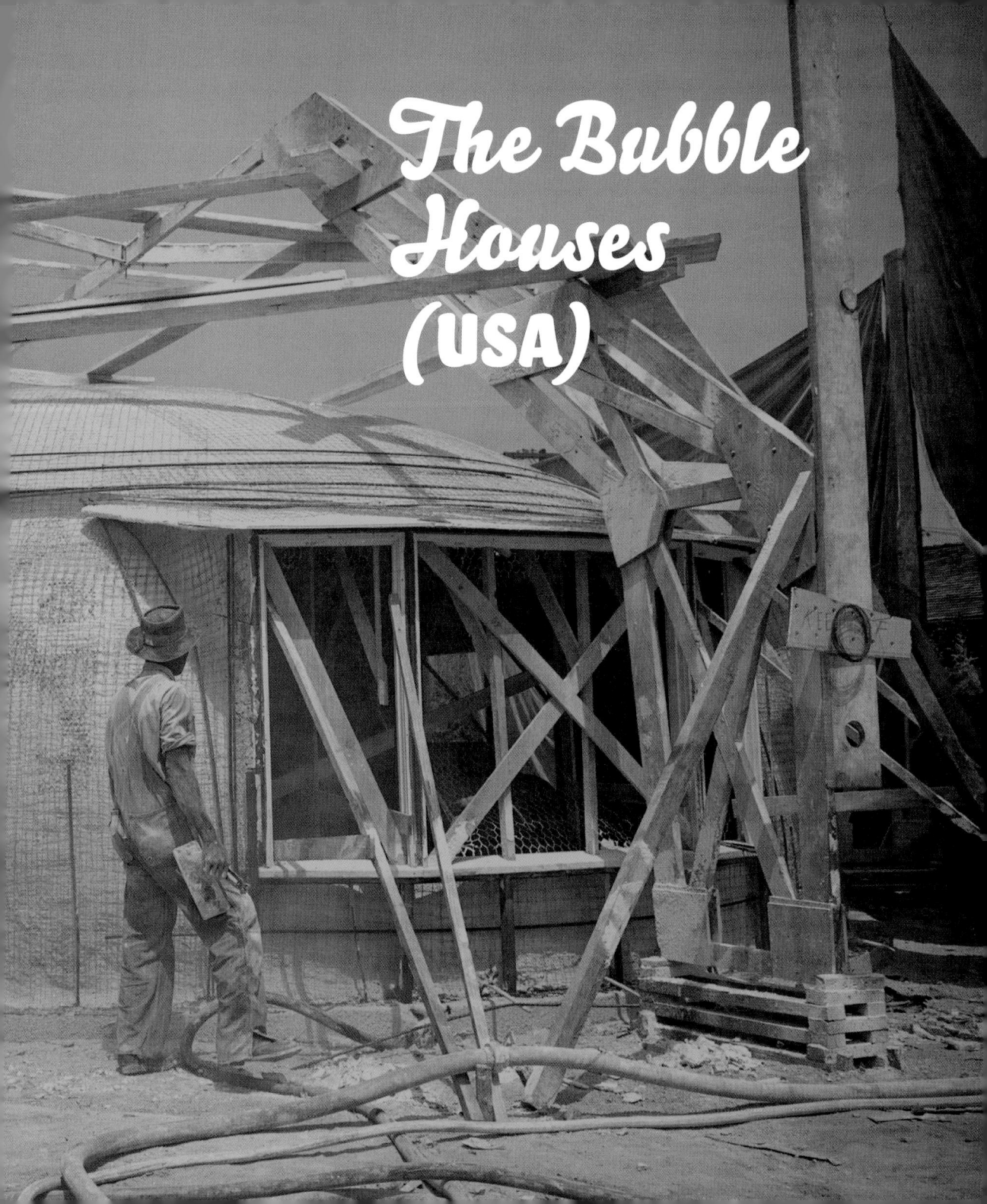

Falls Church
VIRGINIA

opposite top
Fig. 1: Vistors at the Falls Church bubble houses, 1942

opposite left
Fig. 2: A double-bubble Airform home

opposite right
Fig. 3: Neff, with an Airform scale model in hand

Neff's first project using the Airform technique was the completion of twelve houses in Falls Church, Virginia, in May 1942. The ten double-bubble houses and two single-bubble units were built as an experiment to determine the potential for building more, specifically for low-cost defense housing. [Figs.1+2]

    Neff traveled to Washington, DC, in January 1941 to seek government support and funding. He received approval from Jesse Jones, U.S. Secretary of Commerce, who endorsed the project and its financing through the Defense Housing Corporation (DHC). Jones supported Neff's pending patent application and wrote that it was "of importance to the public interest" to develop the Airform houses, since it was possible to construct them quickly and without critical wartime materials.[1] During construction, Jones, who posed at the site holding a gunite nozzle, swung an axe at the side of one of the completed bubble structures to demonstrate its strength (it remained undamaged). It cost a total of $91,000 (approximately $1.4 million today) to build these twelve houses, including $22,000 that was spent on land. The estimated cost was $3,000 for a single bubble and $6,300 for a double. [Figs. 3+4]

FALLS CHURCH

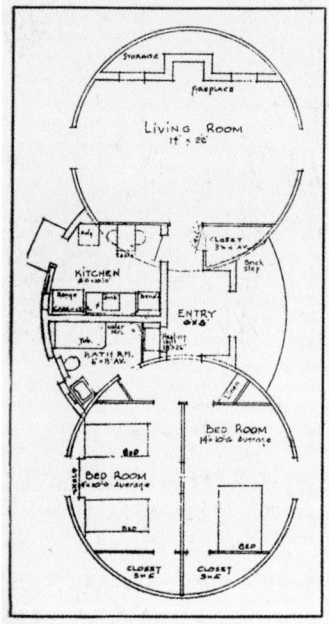

left
Fig. 4: A double-bubble Airform floor plan

right
Fig. 5: An aerial view of the first Falls Church bubble houses

The government officials considered Neff's bubble houses an innovative and inexpensive method for constructing houses for defense workers, particularly in areas where war production was concentrated. In the area of Long Beach, California, there were plans for thousands of houses for aircraft workers at both the Vultee and Douglas Aircraft plants. In each case, more traditional construction was selected instead. Neff did not receive any other commissions from the DHC, and the bubble houses in Falls Church would remain the agency's most experimental, smallest, and permanent housing project.

Neff did not start construction of the Falls Church bubble houses, what eventually become known as "Igloo Village" among residents and locals, until October 1941. The houses were built on a wooded fifteen-acre property approximately ten miles west of Washington, DC, each on a lot with one hundred feet of frontage. [**Figs. 5+6**]

While the houses were under construction, the Goodyear Tire & Rubber Company promoted its development of the Airform balloon with advertisements in both *Newsweek* and *Time* magazines.[2] Combined with various press stories, Jack Linforth, a Goodyear Vice

Fig. 6: The Airform "Igloo Village," Falls Church, Virginia

President, wrote to Neff in 1941, saying he had personally described the Airform development to Henry Ford.

The houses were built using a single twenty-two-foot, eight-inch-diameter balloon from the Goodyear Tire & Rubber Company at an approximate cost of $900.[3] The double-bubble houses were approximately 1,040 square feet and the single-bubble units were 480 square feet. The ceiling height was eleven feet in all the houses. In the double-bubbles, the living room was in one bubble and the two bedrooms, each 10' x 14', in the other. The two forms, about six feet apart, were connected to a traditionally built cinder block structure, with a flat roof that contained the kitchen, bathroom, entrance way, and gas furnace. Vernon Case, the contractor, commented at the time of construction that the houses "could be produced at the rate of 100 in 60 days or 200 in 90 days, using only four balloon forms in the construction cycle."[4] Neff believed the same balloon could be used to build 1,000 houses.[5] All told, each of these original single-bubble houses required approximately 150 sacks of cement.[6] Each house took about two days to finish. [**Figs. 7–16**]

The floors were made of an asphalt tile and finished with a wood floor, which many covered with linoleum. Although fireplaces and bookcases were planned for the units, they were not included in every house. For the exteriors, white paint was mixed with the gunite to create efficiency and reduced cost. For the metal sash windows, Neff chose the same dark green color for the painted wood shutters that he had used on the mansion he built for Hollywood legend and his one-time business partner, King Vidor. Neff asked Mary Lee, the wife

FALLS CHURCH

Figs. 7–11: Constructing the cement anchor and positioning the Airform balloon on top.

Figs. 12–16: Attached, the Airform balloon was inflated on the anchor under constant air pressure. A layer of gunite was sprayed over the balloon, wrapped in a wire mesh, and covered with a second layer of gunite.

FALLS CHURCH

Fig. 17: Cars lined up to view the completed Falls Church Airforms.

opposite
Fig. 18: A Falls Church Airform featured in an advertisement for the Lehigh Portland Cement Company

of Douglas Fairbanks Jr., to be the consultant for the interior design. She made several suggestions for the décor of the first completed bubble house that included traditional furnishings, such as a Duncan Phyfe dining room table.[7]

Neff expected to receive multiple orders from the government and commented that "the tests were spectacularly successful" with the Falls Church bubble houses.[8] He proposed construction of airplane hangars, barracks, bomb shelters, munitions storage, gas and oil storage tanks, and even concrete barges.[9] Despite expressing interest, the government committed to very few projects, limited to several administrative buildings for the Navy in San Pedro and for the Marine Corps in San Diego.

Throughout their construction and after their completion in 1942, local newspapers printed stories and photographs of the "Igloo Village." They reported, for example, on the traffic caused by crowds as large as five thousand people who drove through to see the houses.[10] The unconventional shape remained a curiosity that attracted a steady number of visitors to the site. For some people, the houses appeared primitive in their unadorned functionalism; while for others, like architecture writer Douglas Haskell, the bubble houses were a counterpoint to the glass and steel structures of the modernist aesthetic.[11] Other published responses to the houses said, it "may or may not be 'architecture,'" and it was "like something out of a fairy tale...[a] charming, graceful, completely original and uniquely modern concrete cottage."[12] Another stated, "an odd little colony of Walt Disney mushrooms has sprung up under the trees on Horseshoe Hill, Fairfax County."[13] One reporter alluded to Jonathan Swift's *Gulliver's Travels*, and wrote, "Neff has taken the melon for his model and literally blown it up to Brobdingnagian proportions."[14] [**Figs. 17+18**]

The DHC sold the Falls Church bubble houses in 1947. Residents attempted to buy their individual units, but the lots were sold as a group for $48,000 to a Captain Hampton E. Turner as an investment.[15] In 1961, a developer demolished the bubble houses and built an apartment complex.

It's like Magic
THE WAY THESE HOUSES GROW!

DEFENSE EXPERIMENTAL HOUSING, Falls Church, Va. • OWNER: Defense Homes Corp. of RFC.
INVENTED by Wallace Neff, Architect, Cal. • CONTRACTORS: Case Construction Co., San Pedro, Cal.

Who ever thought the time would come when they'd build a house around a balloon?

Pondering the need for quick, inexpensive yet comfortable emergency housing, a Western architect created a brand-new way to build. Why not inflate a balloon, set doors and windows in place, shoot concrete on the balloon with a cement gun; and when the concrete is hard, deflate and withdraw the balloon?

As you can see, his dream came true. His "bubble" homes are not only practical and attractive, they're fire-and-termite-proof. And — timely thought — they're tough enough to ward off bomb splinters.

Three types of Lehigh Cement were used in this novel experiment in housing — Lehigh Normal, Lehigh Mortar and Lehigh Early Strength. The latter, making service-strength concrete in ⅛ to ⅓ the normal time, helped provide the speed-construction called for by the nature of the job.

When time is short, Lehigh Early Strength Cement *is the* cement to use. When time is plentiful, *it still is* — for its contribution to efficiency and economy is as notable as its speed.

Lehigh
★ **EARLY STRENGTH CEMENT** ★
for **service strength** concrete in a **hurry!**

LEHIGH PORTLAND CEMENT COMPANY ★ ALLENTOWN, PA. ★ CHICAGO, ILL. ★ SPOKANE, WASH.

FALLS CHURCH

Litchfield Park
ARIZONA

opposite top
Fig. 19: Grain bins, one of several proposed uses for Airform construction

opposite middle left
Fig. 20: The construction of an Airform grain storage bin. Neff stands at right.

opposite middle right
Fig. 21: The grain storage bins in use

opposite bottom
Fig. 22: Completed grain storage bins

Coming off a successful project for Cole of California in 1942, Neff built a series of storage bins later in the year for the Southwest Cotton Company (which was eventually bought by Goodyear and that included an adjoining corporate retreat area which the company later renamed the Wigwam.) in Litchfield Park, Arizona. The company used cotton in the manufacture of rubber tires for cars, trucks, and inflatable products such as basketballs and air mattresses—in addition to several models of Neff's Airform balloons.

In addition to cotton, the Southwest Cotton Company had a vested interested in barley, corn, and wheat that was also grown in the area. For grain storage, twenty bins were built approximately thirty feet in diameter and fourteen feet high. Instead of using the gunite process, inexpensive local labor manually applied the cement with trowels. The approximate cost was $1,100 per 200-ton bin, which offered significant savings compared to traditional storage construction.[1] Each bin ensured the quality of stored grains and protected it against varying weather conditions and moisture. [**Figs. 19–22**]

LITCHFIELD PARK

below
Fig. 23: Construction of a double-bubble Airform

opposite top
Fig. 24: A newly completed double-bubble house, 1942

opposite bottom
Fig. 25: Airform houses along the golf course's fairway

In 1942, while the storage containers were under construction Neff secured a contract with Paul Litchfield, the developer of the Wigwam. The three single- and one double-bubble houses provided a contrast to the existing Southwestern architecture on the property. Since Litchfield Park was a Goodyear company town, the bubble houses at the Wigwam may have been a public relations decision by Litchfield. The houses were on the fairway of the first hole of the Wigwam's golf course. Vernon Case, who had helped build the Falls Church "Igloo Village," also assisted on the Wigwam houses in Litchfield Park, Arizona which Neff referred to as the "Desert Colony."[2] The Wigwam eventually became a public resort that continues to operate today. [Figs. 23 – 25]

The houses were published in both architectural and popular magazines. One reporter remarked about the "no fuss, truss— or corners to sweep" ease of housekeeping, an idea that was not previously highlighted.[3] The finished interior was an egg-shell white with draperies and furniture in "soft pastel desert colors."[4] [Figs. 26 – 29] The exterior was painted white with the window sash and trim originally painted dark green. Neff used the same green for the window shutters on the Falls Church houses.

THE BUBBLE HOUSES (USA)

Figs. 26 + 27: An Airform living room

opposite
Fig. 28: Kitchen

THE BUBBLE HOUSES (USA)

opposite
Fig. 29: An Airform residence with garden

Between 1941 and 1942 Neff developed two types of bubble houses for Litchfield Park: Types A and F. The Type A House, a 720-square-foot double-bubble plan, had a living room and kitchen in one bubble and two bedrooms and a bathroom in the other. At the time, it cost $2,500.[5] The Type A had a brick fireplace and a carport connected by a pergola. [**Fig. 30+31**] The Type F was the single-bubble house constructed at the Wigwam and configured the same as the two single-bubble houses in Falls Church, Virginia, earlier in the year. Neff designed a third type that was never built, the Type E, a variation of the Type A plan with a fanned pergola overhang to cool the house. Neff also specified a fireplace in the living room, a glassed-in porch, and a Murphy Cabranette kitchen (a combined single unit with a four-burner stove, oven, refrigerator, and sink). Another unbuilt version of this floor plan called for Plexiglas windows, three skylights, and an overhanging area with enclosed space for a bathroom, with the garage set in back, between the two bubble forms.

Both the Type A and F houses were initially used to host visiting Goodyear guests and employees, and later became available to Wigwam employees during the off-season. Linda Lamm, who grew up in Litchfield Park, lived in one of the bubble houses for a short time with her family. She recalled how, "as kids we thought of the bubble houses as Hostess cupcake kinds of things, or something out of the movie, *Forbidden Planet* [1956]."[6] In the 1950s the Wigwam advertised the "Famous Bubble Houses" in its brochures; the listed prices for the air-conditioned, two-bedroom house, with meals included: $42 to $60 a night.[7]

The bubble houses remained in use through the 1970s, but in the mid-1980s, Goodyear Farms was forced to sell the Wigwam and the houses were demolished, yielding their site to more conventional-looking guest cottages. The grain storage bins were demolished during the same period.

Fig. 30: An Airform residence with garden

opposite
Fig. 31: Detail of an Airform porch with pergola

Loyola University
CALIFORNIA

Fig. 32: A triple-bubble Airform

opposite top
Fig. 33: Engineering students at work in the triple-bubble Airform

opposite left
Fig. 34: Loyola's ROTC cadets with the triple-bubble Airform in the background

opposite right
Fig. 35: The interior of the triple-bubble Airform

Neff built four Airform structures in 1944 for Loyola University (now Loyolya Marymount) in Los Angeles. One, a single-bubble unit with six separate bedrooms and a bathroom, was initially a dormitory for university employees. The three other structures were built together to form a single coherent unit, referred to as the "Triple Igloo."[1] The horizontal grouping of the bubbles is the only known instance where Neff chose to overlap the forms rather than to create passageways or hallways between units, as he did at Falls Church and subsequent projects. It may have been an experiment to test the limits of a single-story traditional dome structure.

The Loyola Airforms were a mixture of cement and gypsum. The exterior surface was treated with a bituminous water emulsion and covered with white paint. In subsequent bubble construction, Neff preferred Portland cement, when available, and a cement wash with a waterproofing mixture for coating the exterior. Neff's design included his signature pergolas, similar to those for the Litchfield Park bubble houses. Eucalyptus bows were typically specified while the classical columns supporting the pergolas were concrete-filled clay pipes covered with stucco.[2] [**Fig. 32**]

For a short time after the war it became the athletic office. It became the headquarters for the Loyola Air ROTC after the program was established in 1948, and continued to be so through the 1950s. The single bubble was converted to a kitchen facility. [**Figs. 33–35**]

The triple-bubble form was demolished in 1958 for the construction of the new campus library. In the early 1960s the single bubble was also demolished and replaced by green space.

Pacific Linen Supply Co.
CALIFORNIA

Completed in 1944, the bubble structure for the Pacific Linen Supply Company, located in the Garment District (Fashion District) of downtown Los Angeles, was the largest Airform that Neff designed (more than one-and-a-half times the size of a basketball court). It measured one hundred feet in diameter and was approximately thirty-two feet at its peak but required no interior support columns, girders, or beams. It offered 7,850 square feet of floor space with five entries and doubled capacity for the company's marking and sorting laundry supply service. Neff placed a circular six-foot vent at the top of the ceiling, and spaced six, six-foot diameter skylights near the top of the Airform's approximately 13,500 square feet of surface area. Additional lighting came from four rows of five large fluorescent lights. Including all materials, the Pacific Linen Airform cost about $70,000 (today, approximately $875,000). Neff's client, R. C. Merritt of the Pacific Linen Supply Company, paid him the equivalent of only $3,000—a transaction that Neff wished to keep secret.[1] [Fig. 36]

Fig. 36: The scaffolding for a 100-foot-diameter Airform

The balloon for the Pacific Linen Company was made of sixty segments, a total of 1,020 yards of fabric, and was anchored to a twelve-inch-thick concrete slab foundation. In the center, the form was tied to a temporary pole for reinforcing. Several air compressors were needed to keep the inflated Airform intact. Although wire mesh and steel bars reinforced the structure, Neff noted that it was only about 7.8 percent of the steel used in traditional reinforced concrete.[2] In addition to its size, the building was unique because the gunite was applied from the bottom up instead of top down. This helped to insure the final shape of the building and to satisfy potential engineering issues if the building were to collapse during construction, which is exactly what happened.

Fig. 37: Preparing the balloon for a 100-foot-diameter Airform structure

opposite
Fig. 38: The scaffolding for a 100-foot-diameter Airform

The balloon collapsed during construction, attracting negative attention in the press and overshadowing Neff's innovative and unusual design. One story in the *The Engineering News Record* reported that, "about 20 men working on the scaffolding escaped death or serious injury as the framework, although shaking violently, remained upright," and speculated, "that a loose piece of scaffold might have crashed through the mortar shell and pierced the canvas, allowing the air pressure to drop."[3] Although the accident occurred during the preparation of the second layer of gunite, it was not accurately reported in the press. After the building was finished, Neff wrote to *Engineering News Record*, tracing the cause of the accident to rotted airproofing material inside the balloon. He emphasized that only the balloon had collapsed, not the dome.[4] [**Figs. 37 – 39**]

By the late 1940s Pacific Linen Supply was no longer in business, and the building was demolished several years later.

THE BUBBLE HOUSES (USA)

PACIFIC LINEN SUPPLY CO.

Fig. 39: Construction of a 100-foot-diameter Airform

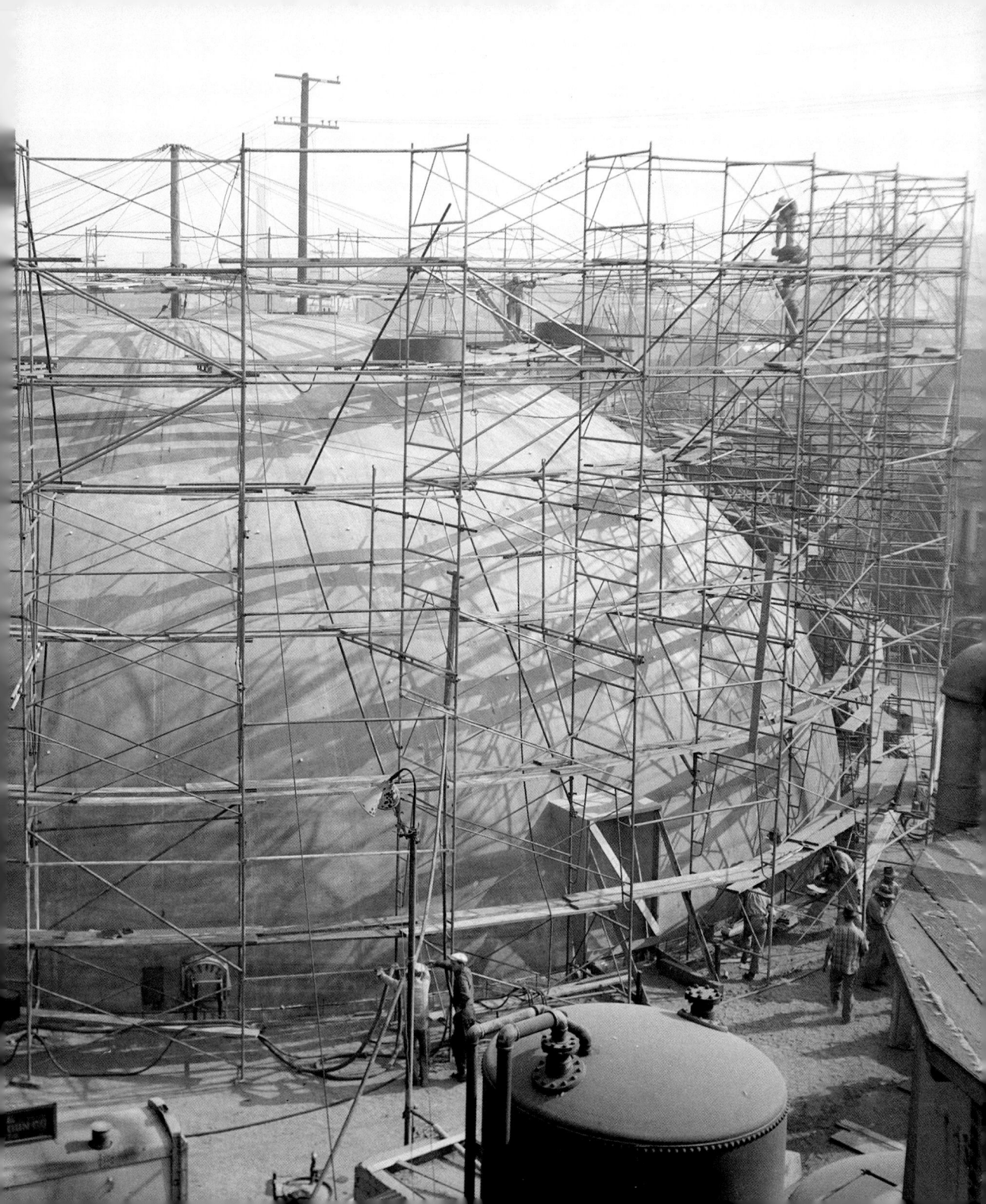

Andrew Neff House
CALIFORNIA

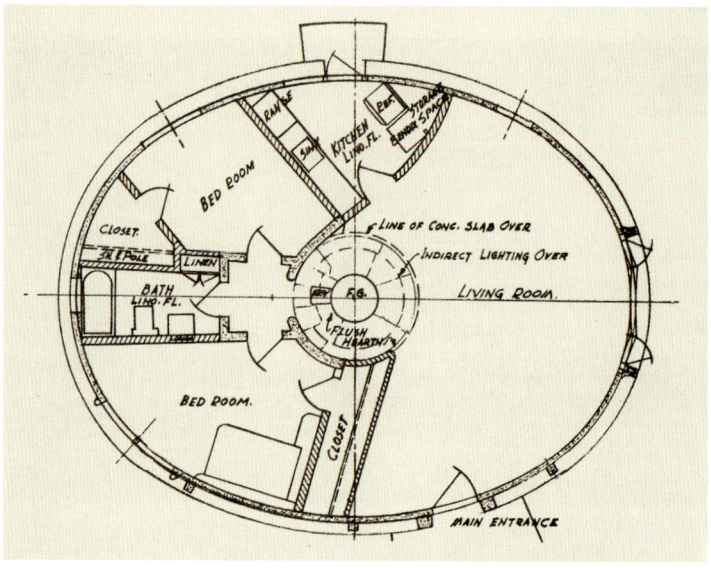

left
Fig. 40: Airforms were featured on the cover of the *Los Angeles Times*'s *Home* magazine, 1947.

right
Fig. 41: A floor plan for a single-bubble Airform residence

The bubble house in Pasadena, California, was built in 1946 for Neff's brother, Andrew. [**Fig. 40**] It features shortened seven-foot interior walls in the kitchen, shared bedroom, and living room so that the space was open to the full height of the twelve-foot dome ceiling, making it feel larger. [**Fig. 41**] A traditionally framed, two-car detached garage and workshop was included and a sizable bomb shelter was built in the 1960s. The house was furnished with modern pieces of the era that included furniture by Russel Wright and Van Keppel-Green. [**Figs. 42 + 43**]

Figs. 42 + 43: The Andrew Neff House today

ANDREW NEFF HOUSE

Fig. 44: A load test using sandbags

opposite
Fig. 45: Neff's Airforms demonstrated his preference for Dutch doors.

Construction of the house was fraught with challenges. At one point, during the first application of gunite, the form collapsed when a piece of wood scaffolding fell and punctured the wet cement. Years later, Andrew Neff recalled how his brother, "tall, stately Wallace, dressed as always in black wingtip shoes, black suit, white shirt, and unpatterned black tie would walk amid the dusty rubble, unflappable, planning the next pour."[1] For neighbors, the implosion sounded like a bomb. Once the house was completed, neighbors began to protest the design. Neff was obliged to landscape the property to make it less visible. The neighbors continued to object, citing structural concerns, but a Cal Tech load test quieted those concerns.[2] [Figs. 44 + 45]

The Andrew Neff House is unique among extant Airform structures for its bomb shelter, built in 1962 under the side yard. It was intended to serve as a functional model and promote a new use for the Airform design. The underground shelter was built on a 4" slab foundation using the same method of construction as the standard bubble house. Neff described the psychological benefit of the shape, saying it removed "the feeling of being closed in," which was common to rectangular shelters.[3] Fifteen feet below ground, the shelter is accessed through the garage by a wooden ladder. The shelter is approximately 850 cubic feet and designed to accommodate a family of four to five, based on the standards established by the Office of Civil Defense.[4] [Figs. 46 – 50]

ANDREW NEFF HOUSE

THE BUBBLE HOUSES (USA)

opposite
Fig. 46: Original interior circa 1947, with Diego Rivera painting at right

Fig. 47: View today of Airform interior with free-standing fireplace

right
Fig. 48: The kitchen

bottom
Fig. 49: The living room

Fig. 50: The sitting area

ANDREW NEFF HOUSE

South Pasadena
CALIFORNIA

Eleanor Ince, the widow of silent film director Thomas H. Ince, financed this single-bubble house built in South Pasadena, California. Ince never lived in the house but the two had a lasting professional relationship. Neff hoped she would finance the construction of several bubble houses on the two lots making up the South Pasadena property, but by 1946 only one was completed. [Figs. 51–54] The house, built shortly after the Andrew Neff House in Pasadena, was an oval with straight sidewalls and 1,752 square feet of floor space. After Neff and his wife Louise separated, the South Pasadena bubble house became her home along with their children. [Fig. 55]

The house was divided into two halves. One contained the living room, a galley-type kitchen, and fireplace, while the other housed three bedrooms and bathroom. The master bedroom, with walls that ran to the ceiling, was placed between the others with half-height walls. The interior kitchen wall and two walls in the living room were also half-height, which kept the form open. For the entrance to the house, Neff installed a Dutch door.

right
Fig. 51: The construction plan for an Airform residence

bottom left
Fig. 52: An air compressor (center) inflating the Airform balloon

bottom right
Fig. 53: A worker applies the first layer of gunite.

```
-- RECORD OF MATERIALS AND LABOR --
            GORDON CLOUGH JOB.
            360 Alta Vista Dr.
            South Pasadena, Calif.

            32' x 42' oval shell
            1752 sq. ft. surface area --

Gunite equipment moved to site about May 3rd or 4th.

First Coat:

        1/2" thick at top to 1 1/2" at base
        Placed May 15, 1946
        Crew of 5 plus 2 carpenters and one foreman
        1 nozzleman
        1 gunite operator
        1 compressor operator
        2 men mixing
        2 carpenters on door and window assemblies
        1 foreman

        Seven hours guniting time

        Mix:  75 lbs. Foat Keens Cement
              25 lbs. costing plaster
              50 lbs. sand.

        Materials used:
            Fost Keens Cement          117 sks.
            Costing plaster             39  "
            Sand

Second Coat: (making total thickness of Shell 1 1/2"
              at top to 4" at base, not including base
              course.)
        Placing mesh and bars (2 x 2 - 14) 3 days.
        Gunite work started May 21. Number of men --
        Finished May 28, 2:30 P. M.
        Crew of 5 men plus one foreman

        Mix:  100 lbs. Fort Keens Cement
              200 lbs. sand
              (base coat mix: 1 to 3)

        Materials used:
            Fost Keens Cement          245 sks.
            Sand

        May 21,22,23 equipment breakdown 3 hrs.
        May 24 gunited inside window and door reveals.
        May 27 outside base course ( average 4" thick.)
        May 28 finished base course quits at 2:30 P. M.
               Total 40 hours guniting time.

Inside Coat:
```

During construction, an executive with the contractor, Pneumatic Coatings, commented on the free movement of air throughout the building. He also noted that "the walls are practically perpendicular to a point equal with eye level and from that point on curve in gentle symmetry to the height of the ceiling, adding a certain classic charm and healthful element to the room, and thereby eliminate former impractical dead air corner pieces."[1]

In 1970, Louise Neff sold the house and adjoining lot. Architect Richard Tanzmann had seen the house with a "For Sale" sign on its pie-shaped lot while driving in the neighborhood. He made immediate plans to buy the house with his wife Virginia. The Tanzmanns were the first to own the three-bedroom house after the Neffs. They remodeled it with shag carpet, rainbows painted on the steps, and purple walls, and tiled the bathroom with mirrored shards of glass. They painted the interior an earthy beige. Virginia explained, "there were no problems with hanging things on the walls and we placed a sideboard against one of the straight sidewalls."[2] For the couple, the curvature

opposite

Fig. 54: Workmen preparing for the second layer of gunite

Fig. 55: The completed Airform

of the walls was most noticeable in the kitchen area. No structural modifications were made while the Tanzmanns owned the house, although at one point there were plans for a two-story free-standing cube for the adjoining property, accessed through a glass breezeway kind of tunnel.

The Tanzmanns eventually sold the house. The new owners had plans to expand it, cutting a large opening in the side, which left it vulnerable to the earthquake that struck Southern California in October 1987 and caused irreparable damage. It was demolished soon after and replaced with a traditionally styled home. Before it was destroyed, the house had remained unaffected by earthquakes for many years, including the Sylmar earthquake in 1971 that caused extensive damage throughout the Los Angeles area.

SOUTH PASADENA

Hobe Sound
FLORIDA

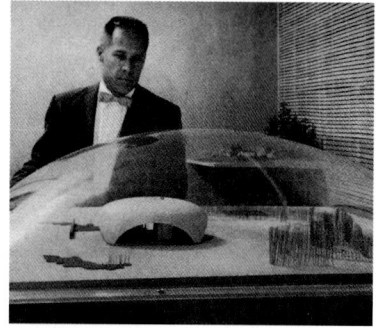

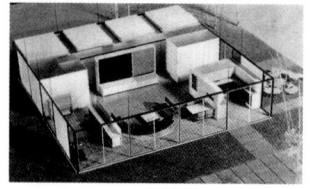

top left
Fig. 56: Eliot Noyes with scale model of his modified Airform structure

top right
Fig. 57: Another scale model by Noyes

above
Fig. 58: An interior scale view of Noyes's modified Airform

Architect and industrial designer Eliot Noyes expressed significant interest in Neff's Airform architecture during the 1950s. He agreed that the Airform's engineered and inventive design had uses in a range of applications, from gas stations and garages to roadside shops and big housing developments. [Figs. 56 – 61]

After continued efforts to develop an Airform project on his own, Noyes received a commission from Joseph V. Reid for the construction of two bubble houses on adjoining lots in Hobe Sound, Florida, in 1953. The Airform design was partially an environmental choice, since it offered protection in areas with seasonal hurricanes and could withstand winds up to 125 miles per hour. Neff estimated the aerodynamic Airform shape was 40 percent stronger against the wind than traditional home construction.[1] Noyes's innovative design modifications and low cost were also factors in the decision to build the houses. [Figs. 60 + 61]

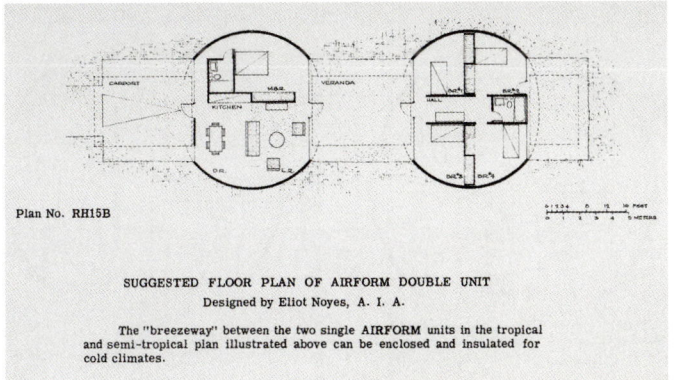

above
Fig. 59: An Eliot Noyes floor plan for a double-bubble Airform

right
Figs. 60 + 61: Illustrations of the interior

HOBE SOUND

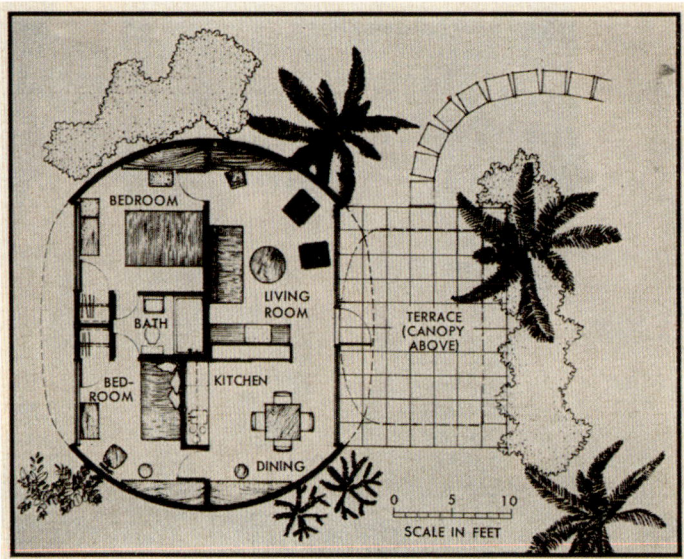

Fig. 62: Floor plan of a completed Airform designed by Noyes

The two Hobe Sound bubble houses, each 600 square feet, were built in 1953 and represented a departure from the fully enclosed bubble structure. In a special agreement between Noyes, Neff, and the AICC, Noyes was granted permission to alter Neff's original design. For the Hobe Sound bubble houses he flattened the dome shape on top and cut two elliptical openings on opposites of the form. He filled them with windows and doors that gave Neff's bubble house a new, open aesthetic. [Fig. 62]

To create the large openings on two sides of the form, a full-size outline was placed on the inflated balloon that indicated the framed area that did not require gunite, leaving the form open. To create the sidewalls of the form, wire mesh was wrapped around the lower portion of the inflated balloon to keep the walls straight. [Figs. 63–65] For insulation, a fiberglass cover was placed between the two layers of concrete.[2] The ceiling was fourteen feet high at the center and interior walls were kept below the curve to keep the plan open and allow for air circulation. [Fig. 66] The structure and process was so unusual, children were let out early from the local school to visit the site.[3] When the house was completed, one report commented, "the Airform house looks like an ice-cream cone without the cone."[4]

left
Fig. 63: A page from an article appearing in Spain featuring a Noyes-designed bubble house

bottom left
Fig. 64: Placing the wire mesh on Airform between layers of gunite

bottom right
Fig. 65: Second application of gunite with Airform opening

Fig. 66: Completed Airform open on two sides

opposite
Fig. 67: Interior decorator Kathryn Smallen seated with Eliot Noyes

HOBE SOUND

above
Fig. 68: Airform interior by decorator Kathryn Smallen

left
Fig. 69: A Noyes Airform at night, Hobe Sound, Florida

Figs. 70 + 71: Additional views of Airform interior by decorator Kathryn Smallen

THE BUBBLE HOUSES (USA)

opposite and below
Figs. 72 – 74: Airform house in Hobe Sound, Florida, today

Only the bathroom was enclosed with a ceiling made of corrugated plastic panels. The kitchenette, closet, bathroom, and utility core were on one side of the house with the living room, eating area, and two bedrooms on the other side. Aluminum verandas were placed on the patios on each side of the house. The large glass-fronted windows added to the cross-breeze that kept the house cool. [**Figs. 67 – 71**]

Today, the bubble houses in Hobe Sound remain intact. The two houses alone costed approximately $3,250 each, approximately $44.00 per square foot today. This included the $4,000 cost of the balloon. Noyes believed the houses could be sold for about $6,500 (not including the cost of land), if they were mass-produced. He estimated eighteen to twenty days to complete each house, with five of those days specifically dedicated to concrete work.

With the completion of the Hobe Sound houses, Noyes was able to change the public's perception of the bubble houses as simple cement igloos. He received widespread attention in various publications such as *Time* and *Life*, in addition to architectural and international press. One review called them, "the most graceful bubble houses yet."[5] Neff also considered the houses an achievement and continued to support Noyes's interest in developing new Airform projects. [**Figs. 72 – 75**]

Fig. 75: A Noyes Airform today, Hobe Sound, Florida

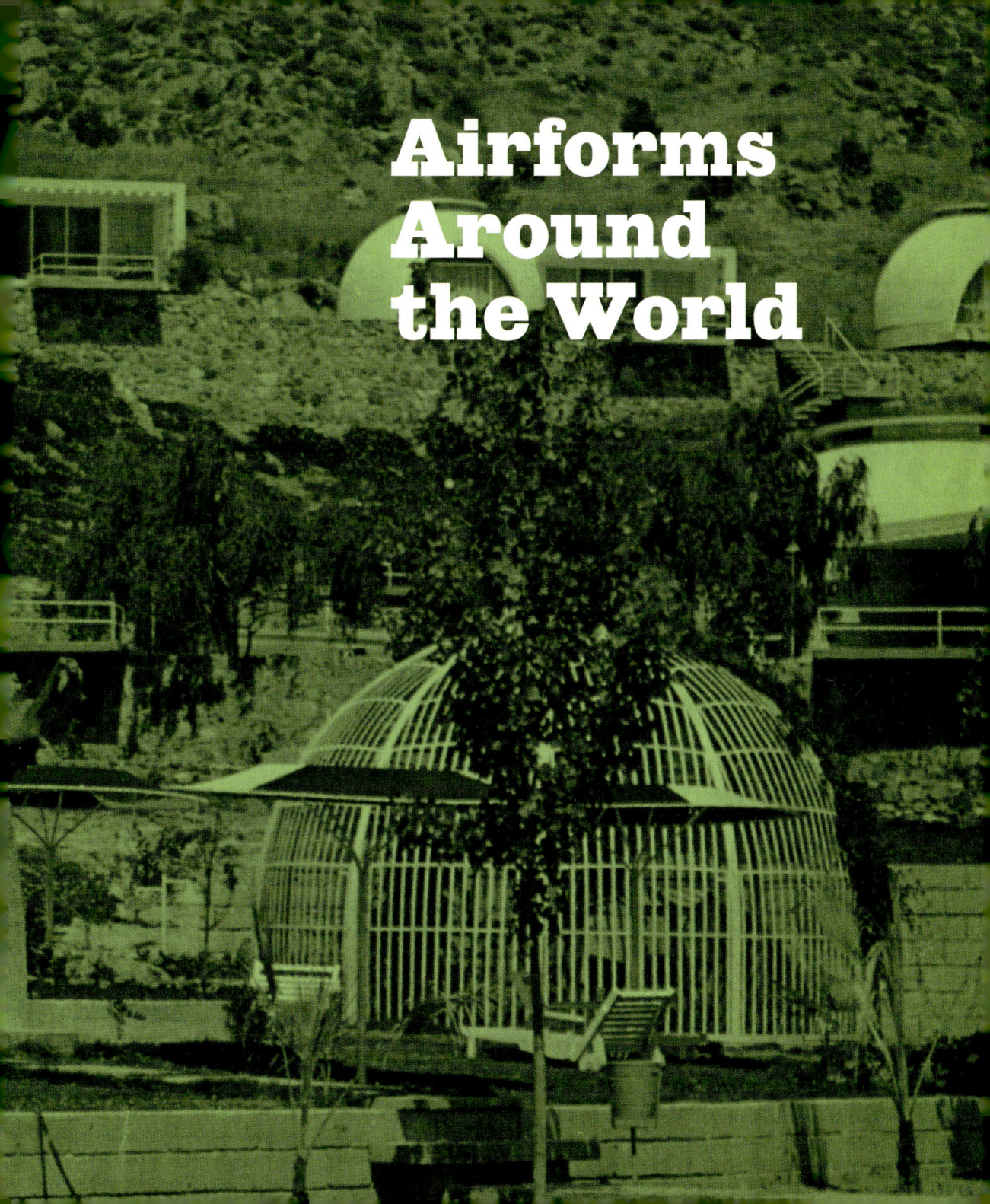

Airforms Around the World

Latin America

Several large-scale Airform projects were planned for Mexico, Central America, and South America during the late 1940s. The flurry of Airform activity in Latin America can be attributed to the efforts of Manuel Reachi. Reachi had become aware of Neff's Airform architecture during his service as a Mexican diplomat and secretary. His belief in the value of Airform construction motivated him to start Construcciones Ultramodernas, S.A., in the early 1950s to promote and build Airform structures.

Neff and Reachi enjoyed a friendship that led to Neff's designing Reachi an Airform home that featured a central 60-foot-diameter Airform, with an atrium-like living room area incorporated into a traditional, rectilinear floor plan. The adobe house, to be built in Ensenada, Mexico, did not get past the blueprint stage because Reachi died in 1955. At the time, two renderings of the house were painted, one by Neff, the other by Elizabeth Calovich, a Los Angeles–based artist. [Figs.1–4]

opposite top
Fig. 1: Neff's painting of the planned Airform residence for Manual Reachi

opposite bottom
Fig. 2: Neff standing with Reachi

right
Fig. 3: Floor plan for the Reachi residence

below
Fig. 4: Artist Elizabeth Calovich made this painting of Reachi's proposed Airform residence.

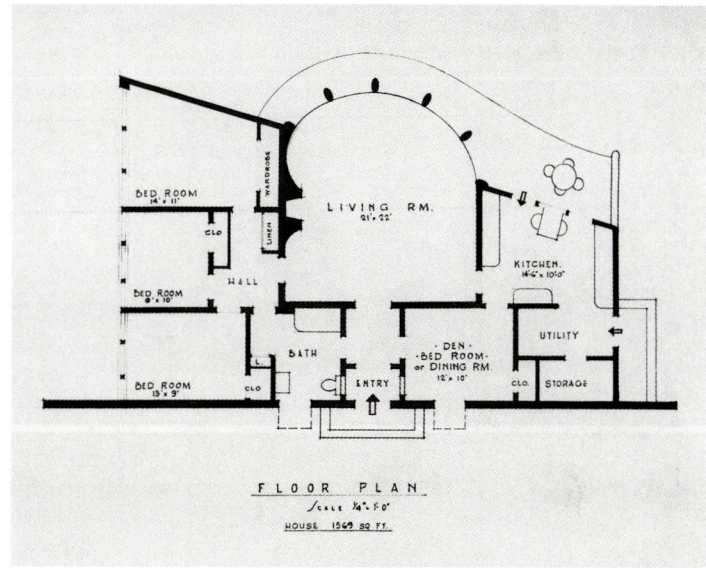

LATIN AMERICA

Fig. 5: An advertisement for Airform licensee in Rio de Janeiro

BRAZIL

Neff found some success building Airform projects in Brazil. The first Airforms in South America were built in Rio de Janeiro and São Paulo. The original 1946 contract with the Brazilian government called for a plan to build one thousand Airform houses each year. By 1947, about three hundred one-bedroom and two hundred two-bedroom houses had been built, in addition to one Airform gasoline station that was built for a naval arsenal. [Figs. 5–7]

Neff did not use gunite for any of the Brazilian Airforms. Instead, he employed a local labor force to lay concrete on the balloon with trowels. At the time, manual labor actually represented a cost savings, since the required concrete mix was cheaper than gunite. Neff recommended trowel application in India, China, and other locations where labor was cheaper.[1] [Figs. 8–18]

Figs. 6 + 7: An Airform gas station, Rio de Janeiro

above
Figs. 8 – 11: Construction showing hand-applied concrete

right
Figs. 12: A brochure promoting Airform housing in Rio de Janeiro

Fig. 13: Completed Airforms, Rio de Janeiro

Fig. 14: A model Airform house, Rio de Janeiro

LATIN AMERICA

| A casa "Airform" é leve e de baixo CUSTO. É uma grande conquista do seculo atual | **MAR** SOCIEDADE ANONIMA AV. NILO PEÇANHA, 12 — Telefone 42-5757 — End. Telegr. "MABRAZ" 10.° andar. - Salas 1013/15 — RIO DE JANEIRO | A casa "BALÃO", feita de concreto, é a unica que resolve o problema da "Casa Popular" |

Consultor Técnico: Arquiteto RAUL PENNA FIRME

Concessionaria exclusiva para o Brasil da

CASA BALÃO

Do Arquiteto WALLACE NEFF, membro do American Institute of Architets de Washington

ESTA É UMA CASA BALÃO

CASA BALÃO DE BAIXO CUSTO

Construção sem armação, de fórma apropriada para resistir à pressão em sua estrutura de concreto.

As casas que vemos acima, são conhecidas como "Casa Balão" e foram produzidas por Wallace Neff, o famoso arquiteto que projetou a majestosa casa dos Pickfor-Fairbanks, em Hollywood.

O arquiteto Neff apresentou planos para casas de um pavimento, com quatro compartimentos, a preços bastante accessiveis, para operários de guerra do R.F.C's Jesse Jones, em 1941. A Corporação de Defesa de Casas autorizou em seguida a construção de diversas dessas casas em Falls Church, no Estado de Virginia.

opposite
Fig. 15: An advertisement for Airform houses in Rio de Janeiro

top
Fig. 16: A completed model home

bottom
Figs. 17 + 18: The interior of an Airform home

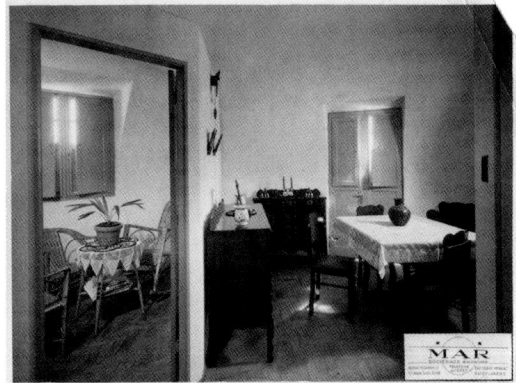

below left
Fig. 19: Airforms in Vera Cruz, Mexico

MEXICO

A handful of Airforms were built in Mexico, limited to four school buildings and twelve houses in Mexico City, and four more in Veracruz, for workers of the Mexico National Railway. Local labor applied the concrete by hand over closed-form balloons. These balloons cost more to manufacture, but were quicker and easier to inflate. The Airforms were completed in 1949. [Figs. 19 – 25]

A restaurant and bar, Los Globos, composed of two forms connected to a flat-roofed building in the middle, was built in Mexico City. Los Globos remained in business through the early 1970s when it became a dance club. Today, such significant modifications have made that there is no indication of Neff's original design. [Figs. 26 – 29]

AIRFORMS AROUND THE WORLD

opposite right, below
Fig. 20: Brochure promoting Airform construction in Mexico

right
Fig. 21: Diagram for construction, Mexico City

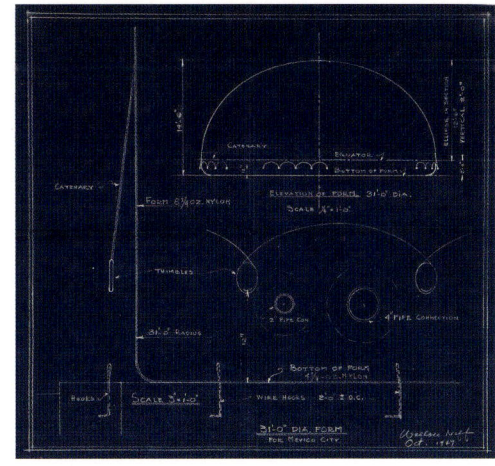

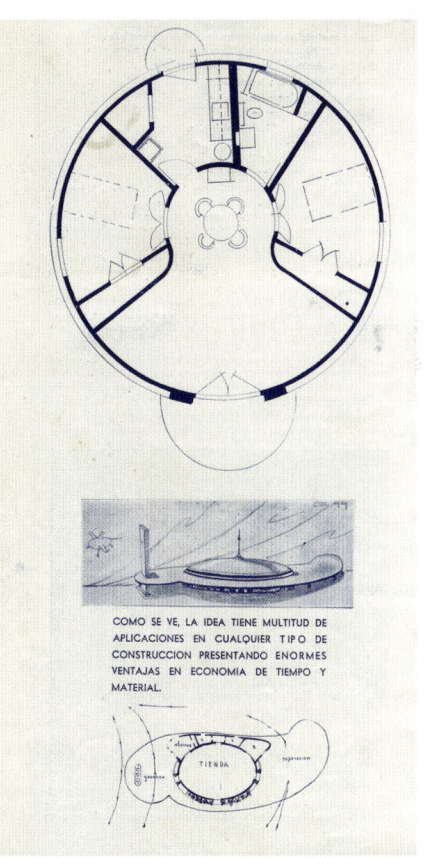

LATIN AMERICA

above
Figs. 22 + 23: An Airform school in Mexico City

right
Fig. 24: Detail of the windows at the school in Mexico City

opposite
Fig. 25: A brochure promoting construction of Airforms for schools

AIRFORMS AROUND THE WORLD

Airform School Construction
MEXICO CITY—29 FT. DIAMETER AIRFORMS USED

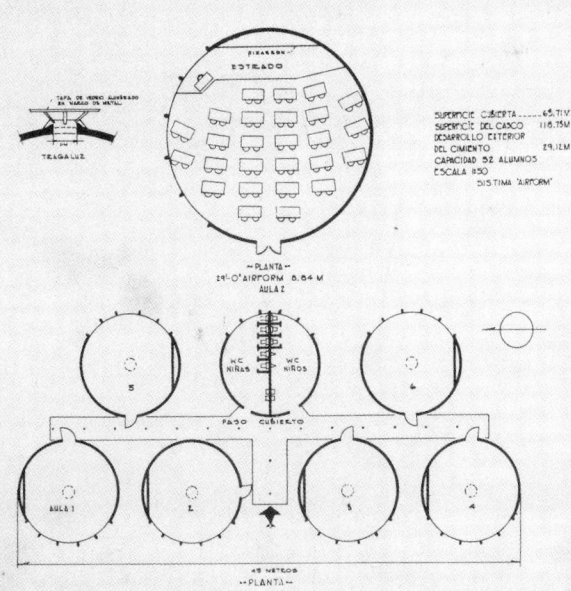

LATIN AMERICA

opposite, this page
Figs. 26 – 29: This Airform was used as a restaurant and bar.

Fig. 30: A rendering of a residential Airform design

NICARAGUA

In 1955, Neff made an agreement with Nicaragua's Compañía Nacional Productora de Cemento. The company was one of the largest cement producers in Central America, with several manufacturing facilities and their own staff of engineers and architects. They constructed the only three Airform buildings in the region, built in the cement yard of the company's Managua plant.

VENEZUELA

In the late 1940s, Venezuela was another large-scale Airform development. A Venezuelan company, Construcciones Aerodinámicas de Venezuela, C.A., contracted Goodyear to manufacture the Airform balloons. When Goodyear did not send the correct ones, Construcciones Aerodinámicas lost the contract. Later, construction of grain storage bins in Venezuela was interrupted by popular revolution in 1948. The U.S. Steel Company and Orinoco Mining became interested in Airform construction in 1952 to house workers along the Orinoco River, but although Neff created several renderings for the project, it was never completed. [Figs. 30 – 32]

right
Fig. 31: Plan for a large residence constructed of multiple Airforms

below
Fig. 32: Rendering of a proposed Airform resort

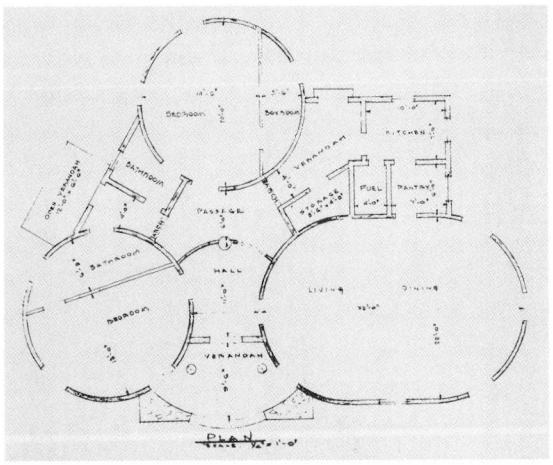

below left
Fig. 33: A brochure featuring Airform resort accomadations

below right
Fig. 34: A set of Airforms that Noyes planned for resort housing

THE CARIBBEAN

After the construction of the two Hobe Sound, Florida, houses, the balloon was sent to Cuba in 1954 where the Constructora Airform de Cuba S.A, a company specifically formed to market and build Airforms, built two bubble houses in Havana.²

Pneumatic International Inc. received interest from several local companies to license and build Airforms in the Bahamas. This included four connected bubble structures, each with separate entrances, for the Pineapple Beach Club in St. Thomas. The combined spaces were used at various times for guest quarters, cabanas, and common space. Their design was comparable in size to Noyes's Hobe Sound bubble houses, except they were open on four sides. The form carried the structural load on its corners, as Noyes had demonstrated in his experimental models. Neff traveled to the site during construction in 1961 and oversaw the new use of a Hypalon paint that offered a fire-resistant protective coating. During one of the renovations of the Pineapple Beach Club resort in the 1980s, the bubble buildings were demolished. [Figs. 33–36]

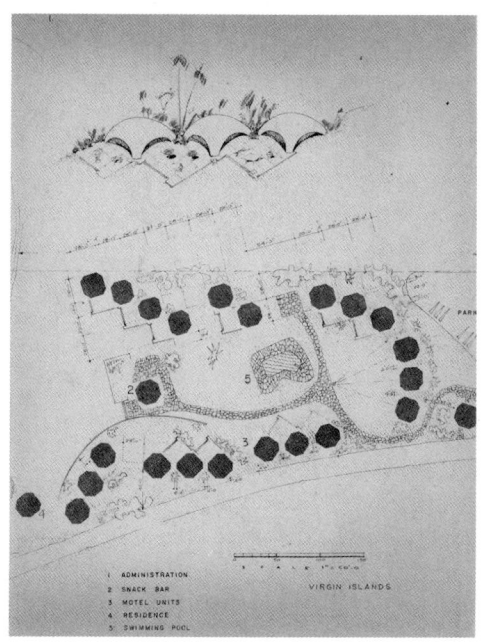

above
Fig. 35: Resort housing in the Virgin Islands

right
Fig. 36: A floor plan by Noyes for resort housing

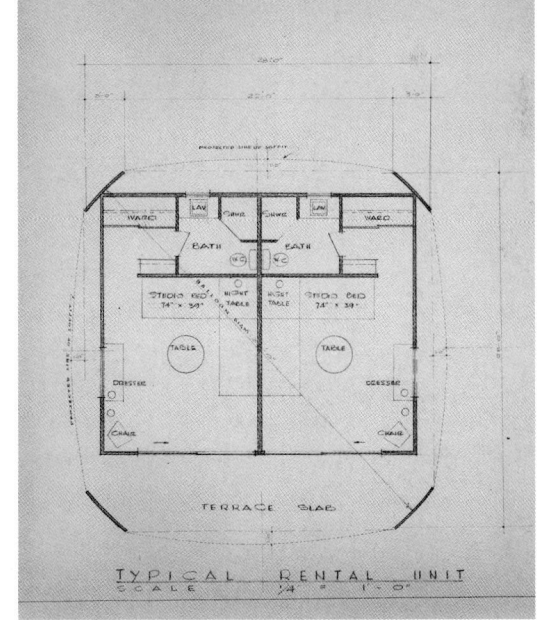

LATIN AMERICA

Europe

opposite, top
Fig. 37: A brochure promoting different uses for the Airform

opposite, bottom
Figs. 38 + 39: Airforms used for wine storage

opposite, right
Fig. 40: Jose Lemos, who ultimately became Neff's Airform licensee in Europe, Africa, and Brazil, seen here in Portugal

PORTUGAL
One of the largest and most successful Airform developments was in Portugal. The original plan called for the construction of five thousand Airform buildings, including single- and double-bubble houses and a large number of storage tanks for wine.[1] Once completed, the Portuguese government considered using Airforms for gasoline stations and ammunition storage, and wanted to create a tourist community with motels, restaurants, and stores. [Figs. 37–39]

After Adolf Waterval's Airform world rights were cancelled in 1956, a Portuguese man named Jose Lemos, a long-time advocate for Airform building, became the exclusive Airform licensee, making good on Waterval's debts and advances. He was responsible for building Airforms in Angola, St. Thomas, and South Africa. [Fig. 40] In 1955, he formed Constructora Iberica Lda. to construct Airforms in the Portuguese-speaking world. Shortly after its formation, the Portuguese government awarded Iberica the entire Airform contract for construction of houses, grain bins, ammunition magazines, and storage tanks.[2]

AIRFORM INDUSTRIAL CONSTRUCTION

CANOPY SHAPES
FOR SERVICE STATIONS, ETC.

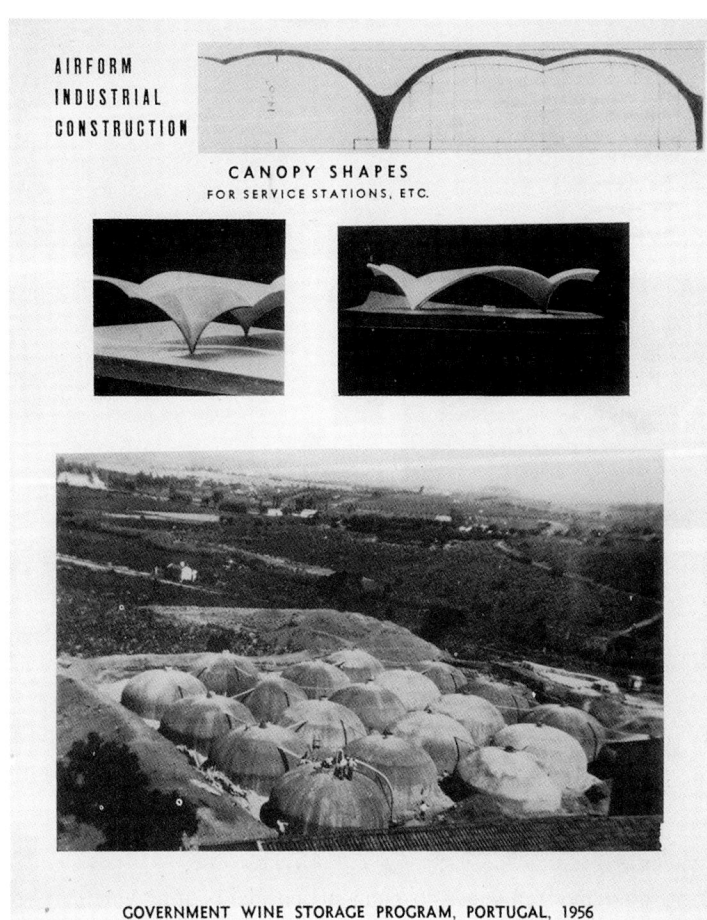

GOVERNMENT WINE STORAGE PROGRAM, PORTUGAL, 1956

EUROPE

Fig. 41: These Airforms were used for wine storage.

The first project was the construction of three Airform houses outside of Lisbon. This was followed by Iberica's development of fifty globe-shaped Airforms built to store wine for government-owned wineries. They were produced at a rate of one every two days; Neff estimated that a conventional tank would have taken at least a month to construct.[3] Testing of the new structures determined the interior temperature was evenly distributed and kept wine safe. At the height of the Portuguese project, in 1958, Neff traveled to Portugal to view construction first-hand and to offer Lemos additional support in his development efforts. According to Neff, tests conducted by French chemists "found that wine stored in these gunite vats was far superior than wine stored in any other container."[4] The success of these storage tanks lead to the government's extensive adoption of Airform construction. [Figs. 41 – 57]

left
Fig. 42: The J.N.V. (Junta Nacional do Vinho) wine facility

right
Fig. 43: Airforms were used for wine storage.

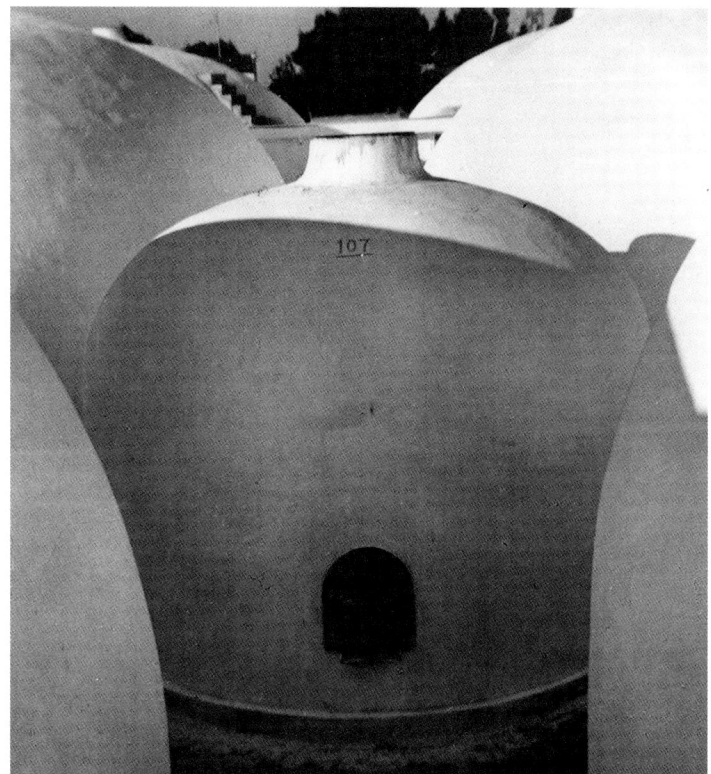

The success of Airform construction in Portugal can be largely attributed to architect Joe A. Wahler of Altadena, California. Wahler was Chief Engineer for the AICC at the time and had worked with Waterval on the project before the his licensing rights were canceled. In Portugal, Wahler taught the local crews about Airform construction and the gunite technique. Ultimately, the difference in experience between Wahler and Waterval added to the difficulties of the project.[5] [Fig. 58]

Fig. 44: Preparation of the foundation for Portuguese liquid storage units

Fig. 45: Positioning the Airform balloon

Fig. 46: Fastening the form to the framework with metal cable

Fig. 47: The air pump inflating the balloon

Fig. 48: The Airform balloon

Fig. 49: Airforms in various stages of construction

Fig. 50: The metal gridwork applied between layers of gunite

Fig. 51: Application of second layer of gunite

Fig. 52: Another view of the second layer of gunite applied to Airform

Fig. 53: Airform balloon carried into position on foundation

112 AIRFORMS AROUND THE WORLD

opposite left
Fig. 54: Metal gridwork was applied between layers of gunite.

opposite middle
Fig. 55: Preparing the site for additional construction

opposite right
Fig. 56: Flat area readied for gunite process between the storage containers

opposite bottom
Fig. 57: Airforms used for wine storage

right
Fig. 58: A Wellington Sears advertisement promoting their industrial fabrics

EUROPE

Africa

above left
Fig. 59: Official opening ceremony for a completed Airform house, near South Johannesburg, South Africa. Neff wrote: "The Mayoress of Johannesburg officially opening the Airform house at Baragwanath."

above right
Fig. 60: An Airform constructed for for French West African military officials in Dakar

opposite top
Fig. 61: The first Airform houses under construction in Dakar, Senegal

opposite bottom
Fig. 62: Aerial view of the Airform houses in Dakar, Senegal

SENEGAL
Between 1948 and 1953 approximately 1,200 Airform houses were built in Dakar, Senegal, with expectations for 3,000 bubble houses a year to replace the indigenous grass shacks.[1] While many belived that the project was financed by the Marshall Plan for postwar Europe, it was more likely funded by the Point Four Program, which funneled development money into Africa and Asia.[2] [Figs. 59–66]

The Dakar bubble houses were mostly single-bubble forms, although many double-bubble homes were also constructed. Some Airform balloons were reused more than a dozen times, reinforcing Neff's view of efficiency and low costs.[3] Construction time was typically less than eighteen hours per house.[4] The cost to build each bubble house was $200 to $300, approximately $2,200 today. This was reduced even further by using pipes to frame the smaller Airforms instead of wood scaffolding, where it was scarce. The houses were considered a success, outlasting traditional construction materials despite characteristic cracks.

AFRICA

AIRFORMS AROUND THE WORLD

opposite top
Fig. 63: Street view, Dakar, Senegal

opposite bottom
Fig. 64: Neff touched up this photograph to enhance the Airforms' perspective and scale.

right, below
Figs. 65 + 66: Exterior views

AIRFORMS AROUND THE WORLD

opposite, below
Figs. 67–73: Dakar Airforms today, with surrounding construction

Urban growth has all but hidden the once readily identifiable bubble houses of Dakar. Few remain, and an even smaller number are in original condition. People have built on to existing bubble houses, adapting them for their own use with toilets, living rooms, and other spaces. The result are walled-in compounds, mixed with traditionally shaped, rectangular cement buildings. One of the original double-bubble houses still stands, housing a restaurant and after-hours bar. [Figs. 67–73]

SOUTH AFRICA

In South Africa, the Airform technique appealed to medical officials in Johannesburg, Durban, Cape Town, and other South African city governments who had condemned the use of ordinary cement blocks as unhygienic and improperly insulated. In 1945, three hundred Airforms were constructed in Johannesburg.[5] In Welkom, a city about 160 miles south of Johannesburg, an additional 50 single-bubble homes were built in 1950. This group was among the first to be built with a ventilator on top to circulate air to keep the interior cool in warmer climates.

EGYPT

Former British Army Colonel L. B. Barrington and the Anglo-Syrian Trading and Construction Company facilitated the construction of two Airform buildings in Cairo in 1949. The first Airform was included in the Société Royale d'Agriculture's XVI Exposition Agricole et Industrielle du Caire as a solution to the city's housing shortage. Neff was awarded a gold medal by King Farouk of Egypt in 1949 for his contribution

In 1964, the Egyptian government reconsidered Airform, intending to make the form square, with a domed roof. Schools in rural areas and community houses were planned, but Neff believed it would be too expensive to adapt the form to public preference.[6] The square-plan project did not advance, although several bubble houses were later built in Ismalia.

ANGOLA

In 1956, a Portuguese venture built four hundred Airform houses for employees of the Bengal Railroad company. Double-bubble houses were also constructed. Today, many of those larger Airforms are schools for young children. [Figs. 74–78]

Figs. 74 – 78: Today, this set of Airform buildings is a school in Angola.

Asia

THE MIDDLE EAST

Under the U.S. State Department's Point Four Program, ten Airform containers were built for storing wheat, barley, and corn in Ruseifa, near Amman, Jordan, in 1953. Each of Jordan's ten storage bins featured straight side-walls and were constructed using the same balloon. Each had capacity for five hundred tons of grain. An conical floor created a slope that increased the efficiency of loading and unloading grain, assisted by a conveyor system that was built at the top of each bin. The bins helped the country stabilize its crop supply and protect against the impact of crop failures. The Airforms served as a model for similar development in Irbid, Karak, and Nablus (then part of Jordan).

In 1945 Neff partnered with Standard Oil of California to construct several houses for their company staff and local labor on their Middle Eastern oil fields. Stores, commissaries, and other buildings were also planned under the auspices of the Arabian American Oil Company (Aramco).

In 1950, L. B. Barrington constructed a licensed set of houses for a resort in Nebioglu, Turkey, in addition to several houses in Kuwait. There were also plans to construct five hundred Airforms in Harper City, Liberia, but only a small number were built for military housing. [Figs. 79 + 80]

right
Fig. 79: An Airform with a ceiling vent

below
Fig. 80: Resort housing in Turkey

ASIA

below left
Fig. 81: A story in the March 1964 *United Nations Review* **mentioned the Airforms in Karachi, Pakistan, and included them on the cover.**

below right
Figs. 82 + 83: Completed Airforms for military housing, Karachi, Pakistan

PAKISTAN

The Pakistani military commissioned thirteen bubble houses in Karachi in 1952. Today the houses remain as living quarters by the Pakistani Navy. Two typologies were constructed: a 22' × 32' ellipse and a 27' square with a cloistered roof. Edward D. Chang from China, a licensed architect in Karachi, served as the Chief Engineer on the project for the AICC. Chang also supervised the application of Insulcrete, an insulating material used in tropical and subtropical locations that was ideal for the Karachi bubble houses. [Figs. 81 – 83]

Figs. 84 + 85: Construction of Airforms for military housing, Karachi, Pakistan

INDIA

The number of bubble houses built in India is unknown, especially since a large number were constructed with an unlicensed balloon, without Neff's awareness or permission. In some cases, they were constructed with hybrid materials and modified designs. Later, a more formal plan to address destitute housing areas was developed for major Indian cities as part of India's Second Five Year Plan in 1959, but no officially sanctioned constructed bubble houses were built at that time. [Figs. 84 + 85]

APPENDIX — A
Interview

An interview with Mary Mayhew and Kathy Miles, the daughters of Mr. and Mrs. Robert J. Carmody, former residents of a double-bubble house in Falls Church, Virginia.[1]

How did you come to live in a bubble house?

MARY: After the Second World War, there was a tremendous housing shortage in the Washington, DC, area. People were on waiting lists to get apartments because there was such an influx of people. Our dad heard about these houses and eventually found this street, Horseshoe Drive, with these igloo houses on it...we lived in one for just under eleven years.

How did your parents like living in the house?

MARY: My dad said he was embarrassed to bring my mother to the house when we came down to see it. Apparently it was pretty rough, unfinished, even though it was solid construction, just this huge dome of concrete. The floor was a slab of concrete with black linoleum tile. Sort of like a basement.

KATHY: And all the houses were painted white on the outside.

right
Fig. 1: Sisters Kathy Miles and Mary Mayhew in front of their childhood home, a double-bubble Airform.

below
Fig. 2: One of the Airform "igloos" in snow-covered Falls Church, Virginia.

How was the interior of your house painted?

KATHY: I remember the living room was green. All green. Up to the ceiling. My father was not handy. He was completely covered with paint. He used to say the only thing he could fix was a sentence. After that he got someone to do the painting for him. My mother ended up doing all the rest of it. If it was going to be done, she did it.

MARY: In the other dome was our room. It was pink, and our parents' room was blue. My mother went through a phase where she painted the hallway different colors. One year it was maroon. I think because she couldn't put up artwork she painted the hallway and the flat ceiling. Another year it was aqua.

She did not hang photographs or art on the walls?

MARY: There was no wallboard. It was concrete, so there wasn't any way to hang pictures or anything. We had a bookcase that Mother used to lean pictures on against the curved wall in the living room. She did prop things up like paintings on various pieces of furniture.

How old were you when you moved into the bubble house?

MARY: I was about two years old.

KATHY: I was five and a half, and started first grade when we moved in. I remember in school we were asked to draw our house. So I tried to explain it, but really couldn't. They thought I was making it up…so that kind of set the scene and got me started on the wrong foot in first grade. It was such an isolated road, a lot of people didn't even know it was there, and hadn't seen the houses.

What do you remember about your house?

MARY: When you walked in through the center door, you were in the hallway that connected the two bubbles. To the right, that whole bubble was the living and dining room. That was big with a lot of furniture. There were no electric plugs—only a couple on the wall; so my mother would run the cords under the floral rug and put plastic protectors where the cords weren't covered, so we wouldn't trip over them. There were lamps and things in the middle of the room. No ceiling fixtures. It was hard to light the room because it was big and there was shade from outside, and the windows were only on the front and back. There were no side windows. The doors were the old solid wood type and the walls were thick.

KATHY: The interior walls were plaster. Wood-framed. The kitchen and bathroom were made with wood framing. Between the two bedrooms, the walls went all the way up to the ceiling. There was no gap. They were flat solid walls.

What about the hallway?

MARY: In the hallway, on the right in the back, was the kitchen and a closet that my mother used as a pantry. If you walked to the left, that bubble was split down the center with a straight flat wall that went all the up to the ceiling. The front half of that dome was the children's room that my sister and I shared. The other room was my parents' bedroom. There was a little bathroom with a tub. It wasn't tiled. It was pretty simple. The bedrooms were big. My sister and I each had a twin bed on opposite sides of the room, and there was quite a distance between us with a play area in the middle, and we had dressers and a closet.

And the kitchen?

MARY: We had a tiny gas cooking stove, and my mother had to stuff things in the oven door to keep it from falling open. We had a white-porcelain sink and a small refrigerator. At some point during the [19]50s, we did get a clothes washer.

KATHY: The kitchen was so tiny.

What about the windows?

MARY: The windows were only on the front and back. There were no side windows. Each window had three sections: they were casements that were painted green. They opened outward, and you could sit on the window sill, and slip in and out.

KATHY: The windows had a wood frame, but there was no other wood, so our mother hung the curtains from concrete nails. There was a valance put in to hide the top of it, just a piece of plywood

Fig. 3: A single-bubble Airform home

covered with fabric. It kind of bowed out and sat there because of the tension inside the window frame. There were screens on the windows.

Did you have a garage or driveway?
MARY: Each house had little flagstones walking up to the front door from the street. There were no driveways. You parked on the street, and when you got out of your car, you walked over the ditch on a concrete slab.
KATHY: They put a pipe in the ditch in front of each house and covered the top with dirt, so you could walk over it to your house from the street. Like a bridge.
MARY: The ditches were made of concrete.
KATHY: As a child the ditch seemed huge but it was probably a foot and a half or two deep on sides of the road for the water to run off.
MARY: They were great for playing. After it rained, Kathy and I would run out, make boats, and send things floating down the ditch.

Fig. 4: Kathy and Mary playing in the "Igloo Village."

What was the street like?
MARY: The street was paved at one time and seemed mostly like gravel.
KATHY: There were no street lights. It was a scary dark place for people who didn't live there. There was a porch light in front, but that was it, and people didn't always put them on all the time. It was a very dark street.

What was it like growing up in the house, in the neighborhood?
MARY: It was a great place for kids. My memories are very happy. It was a lovely setting. There was nothing else there for acres and acres, and you could go out in the woods and play, climb trees.

How was the neighborhood?
MARY: It was just a normal neighborhood with a lot of kids. I remember for Halloween, we'd go to the individual houses, and we'd stay for a while, and the mothers would do things like make candied apples and popcorn. It was such a small community.

Did the neighborhood ever change?
KATHY: In the first four or five years, there was quite a difference in the type of people who lived there compared to the people who moved in after, during the last five years [we were there]. There

was a change in the occupational and educational level, and it became more blue-collar.

KATHY: The rents stayed low.

Your parents rented the house?

MARY: I remember at one point in the [19]50s, my parents paid $50 a month for rent.

KATHY: Later I think they paid $60 a month, then maybe $70.

How would you describe the neighborhood, the area?

MARY: It was all trees when we moved there. I don't know how they built the houses, because there were so many trees. There were two single-bubbles up at the top of the street, which was a gravel kind of road. At the end, there was something we called the turn-around, a round area. Those were occupied by single people, or they didn't have kids.

Did you have a garden?

KATHY: When we moved in there was nothing but dirt around the house, but my father over the years went to a lot of trouble to put in grass by bringing in top soil and having seed put down and had a good lawn which he enjoyed, except for the mowing with a push-mower.

MARY: He took out twenty-two trees in one area just to create a little lawn and my mother planted irises and zinnias. Over the years it was really very sweet.

KATHY: Our father used to call the front lawn a poor man's Riviera…people would sit outside on the lawn in the evening, getting together while the kids played.

Was there any social stigma or novelty about living in a bubble house?

KATHY: The older I got, the more I noticed the stigma about living in an igloo house, and I felt a little more strange about it. A lot of parents wouldn't let their kids come visit us, because they thought we were too weird for living in the houses.

MARY: We realized we were a source of curiosity for people who lived in regular houses. Sometimes kids in school would make fun of us and ask us if we were Eskimos.

KATHY: For years, even as an adult, I never mentioned that I lived in an Igloo.

Did you call them Igloos too?

MARY: Yes. Single-bubble and double-bubble igloos.

KATHY: In the early days it was Igloo Village because everything was a Village then. No one ever called them bubble houses, just igloos.

People visited the neighborhood?

MARY: Yes, on Sundays particularly, a lot of cars would drive up our street with the windows rolled up tight with the people inside looking at us.

KATHY: Even in the summertime, they rolled up the car windows when they drove down the street.

MARY: We used to laugh because it was really obvious. They'd drive slowly; looking at both sides of the street, then turn around and leave. We sort of felt like we were in the zoo, people were curious about the houses.

KATHY: People would come up and point, and we felt a bit like we were on display. They thought we were strange.

MARY: Our father was very cautious, conservative financially about purchasing a house, so we stayed a long time, and my mother was very eager to leave the community. She was raised in the city, but he was a country boy, so he thought it was great. She was a city girl. It was not the house she dreamed of.

Why did you move from the house?

MARY: Our mother desperately wanted to live in a real house, and I wanted a room of my own, and Kathy wanted her own room. My parents bought a brick, three-bedroom, one-bath house in Poplar Heights, and later added another bedroom and bathroom in the basement.

How do you think living the bubble house affected you?

MARY: I think living in the house gave us awareness and made us more tolerant of other people's lifestyles.

KATHY: I think it did make me more tolerant of people who were different. We grew up with a broader perspective.

Do you miss living in the bubble house?

MARY: I live in a house now with cathedral ceilings and windows that look out into trees—and my sister has the same kind of place—and that comes from living in the igloo with high ceilings and growing up in a wooden area.

How comfortable was it for you to live in the house?

MARY: Since the house was flat on the ground, it made it very cold like a basement floor. That was really the only thing that made it unpleasant...that it was damp and chilly. Of course I'm thinking of it, remembering it as a ten-year-old.

Did your house have a fireplace?

MARY: No. There was no air conditioning either. Radiant heat would have made it paradise.

KATHY: There was a chimney, an exhaust for the furnace; It had tar around it and was painted and there was a white, maybe concrete, liner around the black chimney that you could see from outside.

Was the house cold or damp?

MARY: Because it was concrete, the walls and floors were cold. We had little scatter rugs but we didn't have carpeting. I always remember having cold feet. There may have been a wood-burning stove when we first moved in. There was a gas furnace with ductwork. It was drafty. The house across the street had a fireplace. They built a flat wall against the curve in their living room for it.

KATHY: In the summer, it stayed fairly cool, but damp, so that's when the mildew encroached.

What is one of your best memories of living in the house?

MARY: Kathy and I climbed over the top of the house. That was a play area in the nice weather. There was a small cement enclosure for the gas or electric meter at the back of the center hallway outside. It was just enough to create a shelf for us to jump up on, then we'd climb up to the ledge that stuck out on top of the window and crawl up the dome to the top and run around up there in our tennis shoes or bare feet. That was one of our favorite things. I think our parents encouraged us and didn't mind. The center area of the house had a tar and gravel texture, so we didn't climb on that in our bare feet. We would sit on the dome. I would sketch and take the cat up there. Kathy was much better at climbing and scampering around. You could either climb down the same way or climb down the poplar tree that was in the front of the house. We also had a mulberry tree to the right of the right front door that we trained to grow up and over the porch, and there were tons of birds that would feast on the mulberries.

KATHY: I used to read up there during the summers under the tree. It was cool and pretty. Sometimes I sat on the very top. Most of the time, I sat on top of the window ledge and leaned against the bubble. One time I wanted to bring a wooden kitchen chair up there while our parents left us with a babysitter, so I tied a rope to it. Unfortunately, when I was hauling it up, it swung, and I broke the window...and I got in trouble.

APPENDIX — B
Patents of Wallace Neff

As Neff developed and refined his architectural use of a pneumatic form for construction, he realized the value of his innovations, and received U.S. patents to protect his designs and techniques. In 1941, after several years of research, Neff received his first of twelve patents related to Airform construction. In the development of this first patent, Neff cited the Hayden Planetarium, in New York City, as a similar form, although his elaboration in the patent quickly points to the fundamental differences with the Planetarium and other traditionally built dome structures. As he stated in the patent application, the Airform has:

> the provision of a thin-shell concrete structure of the continuous wall or span type which is so constructed as to eliminate the requirement for joists, beams, studding, girders, and like structural parts, and which nevertheless possess sufficient strength as to withstand all normal stresses to which it may be subjected; the provision of thin-shell concrete structures of barrel, vault or dome shape constituted by a continuous wall or span in which all stresses are predominately those of compression, with the shell being substantially self-sustaining and requiring only minimum reinforcement for the negligible lateral stresses.[1]

below and opposite
Figs. 5 + 6: Patent Drawings

Building Construction
Patent Number: 2270229 / Granted: January 20, 1942
Details the specifics of Neff's pneumatic construction process and technique based on a flat-bottom balloon form. Neff referred to the design as form "#1." [Figs. 5 + 6]

Design for a Dwelling
Patent Number: D127276 / Granted: May 20, 1941
A single bubble house with more of a peaked dome shape, and a louvered double-door entry with three matching windows, non-symmetrically placed along the form.

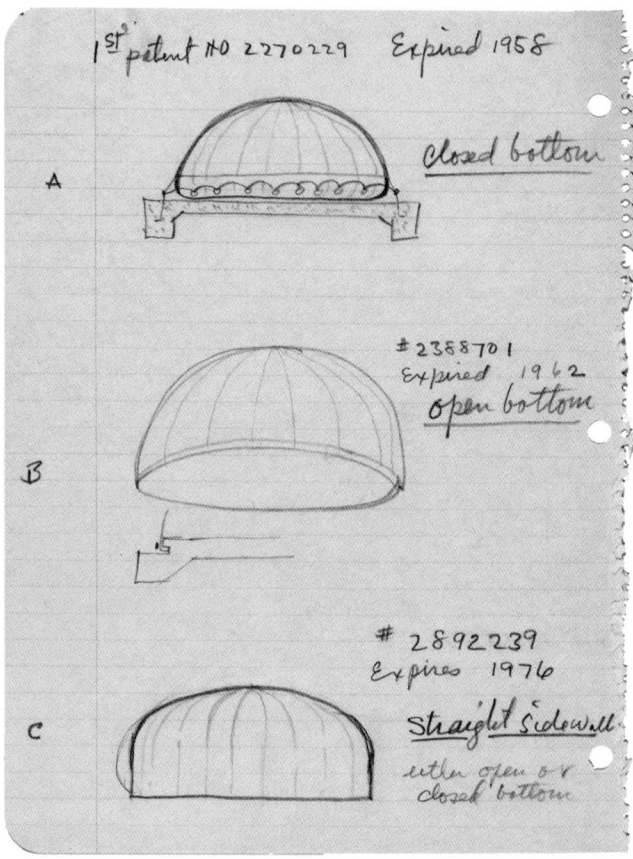

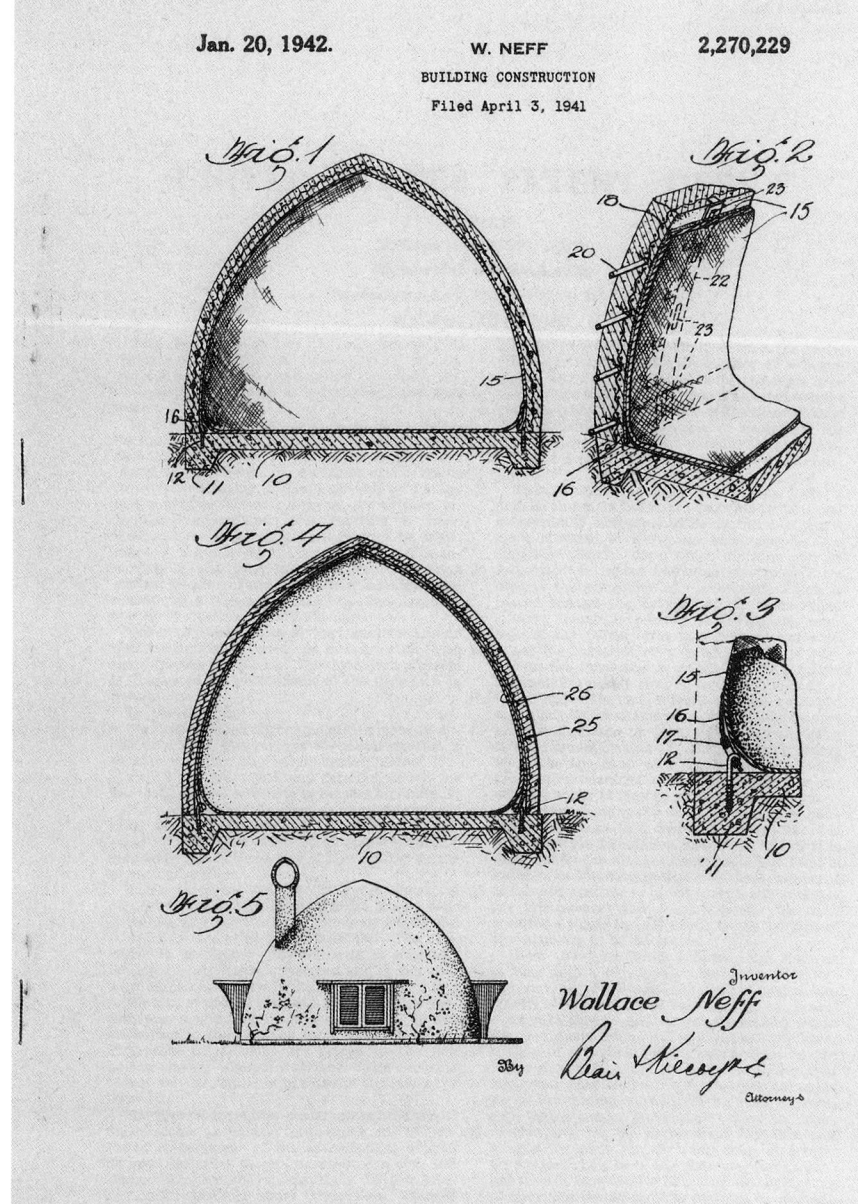

APPENDIX — B

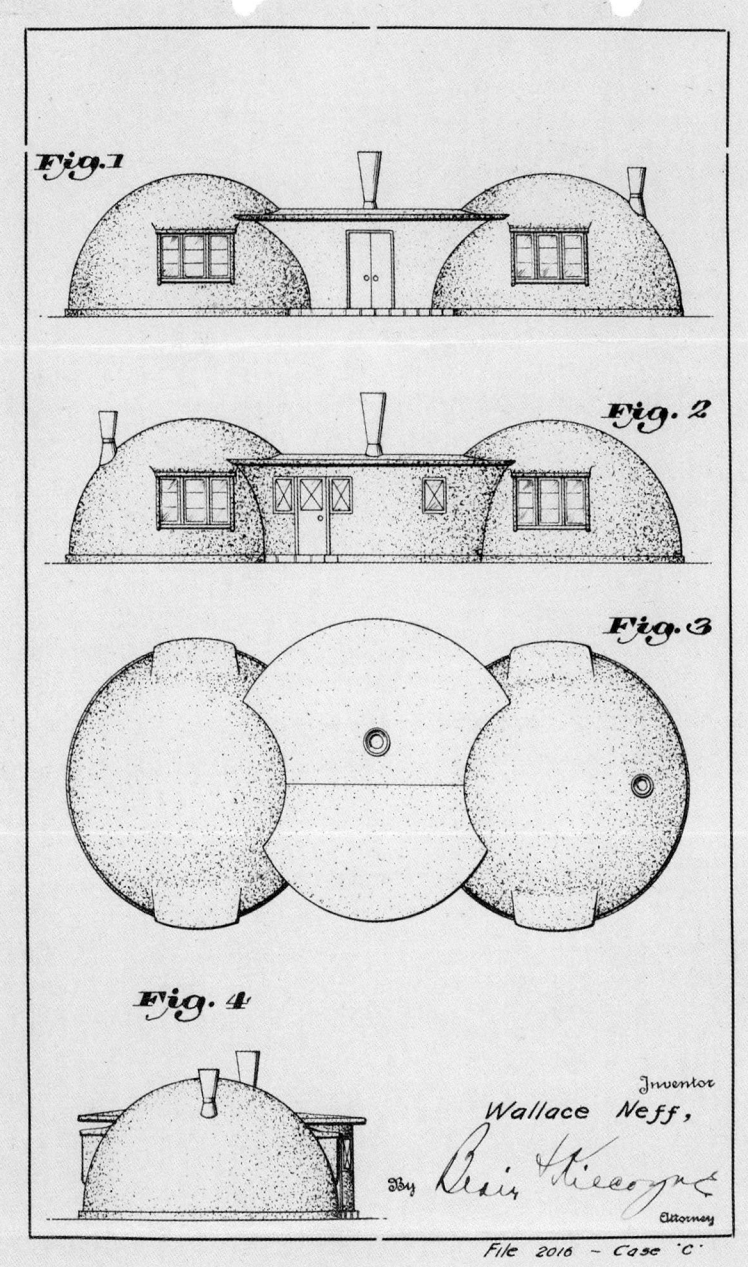

opposite
Fig. 7: One of Neff's patent drawings for a double-bubble house

overleaf, page 142
Fig. 8: Patent drawing

overleaf, page 143
Fig. 9: A patent drawing for a Quonset-type Airform

Design for a Dwelling
Patent Number: D13299 / Granted: July 7, 1942
A barrel-shaped bubble form, similar to a Quonset hut with a louvered double-door entry on one end, and another set of louvered double doors on the side, and two sets of matching windows on the two long sides and a set at the end without the entry, and included a (fireplace) chimney.

Design for a Dwelling
Patent Number: D133658 / Granted: September 1, 1942
A double-bubble house with a connected, flat-roof structure. Distinguishes between the type of front door (single or double) and details the location of the (fireplace) chimney. Windows on two sides of each form. [Fig. 7]

Building Construction
Patent Number: 2335300 / Granted: November 30, 1943
Details the application of gunite (concrete) from the top of the form down. [Fig. 8]

Building Construction
Patent Number: 2365145 / Granted: December 12, 1944
Specifically details construction of Quonset hut-like form. Neff referred to the design as form "#3-a." [Fig. 9]

Design for an Airplane Hangar
Patent Number: D139953 / Granted: January 9, 1945
A low, elongated single bubble form, with no entryway indicated or windows.

Design for a House
Patent Number: D140059 / Granted: January 16, 1945
A single bubble house with a triple French-door entry and a single window on two sides of the form. Design closest to completed Airform built for Andrew Neff in Pasadena, California, and the bubble houses in Litchfield Park, Arizona.

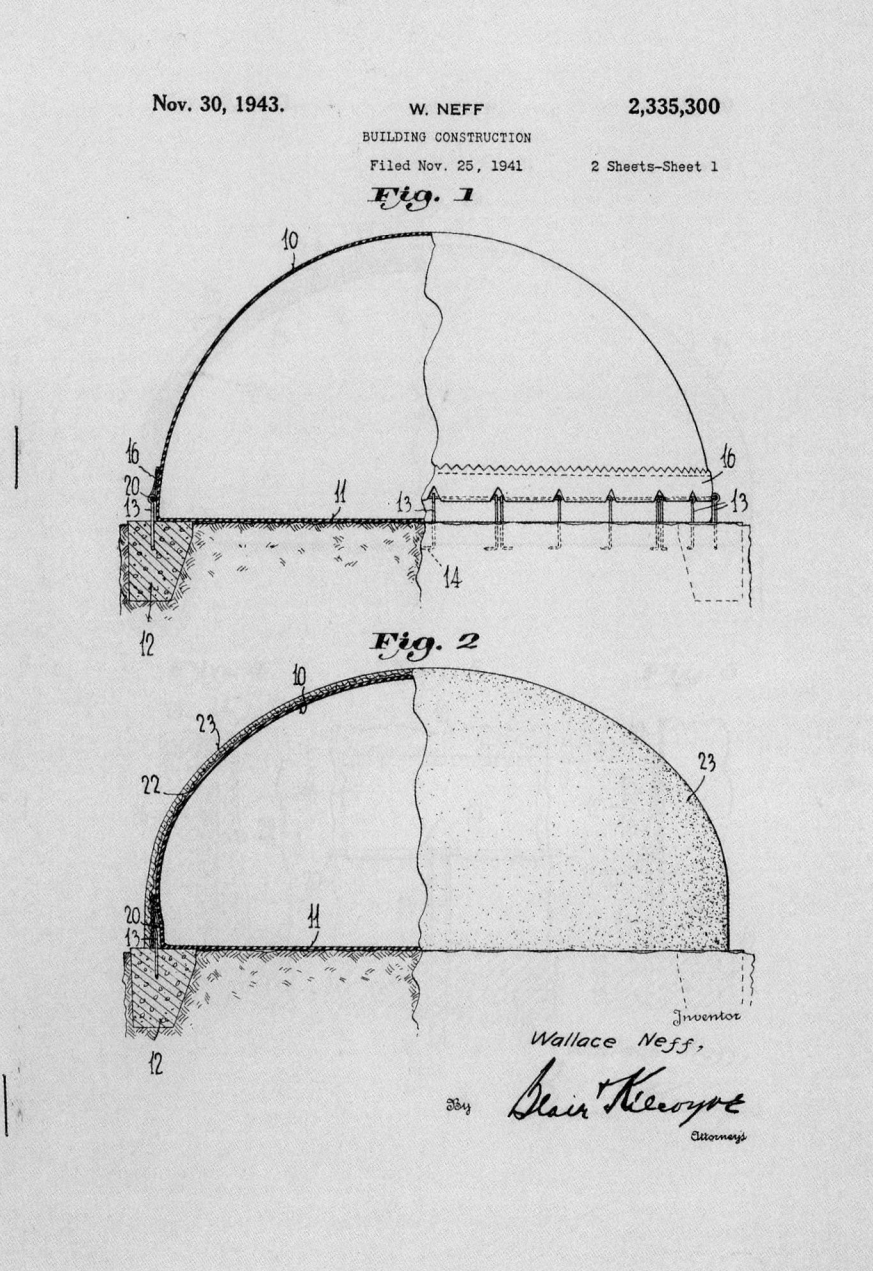

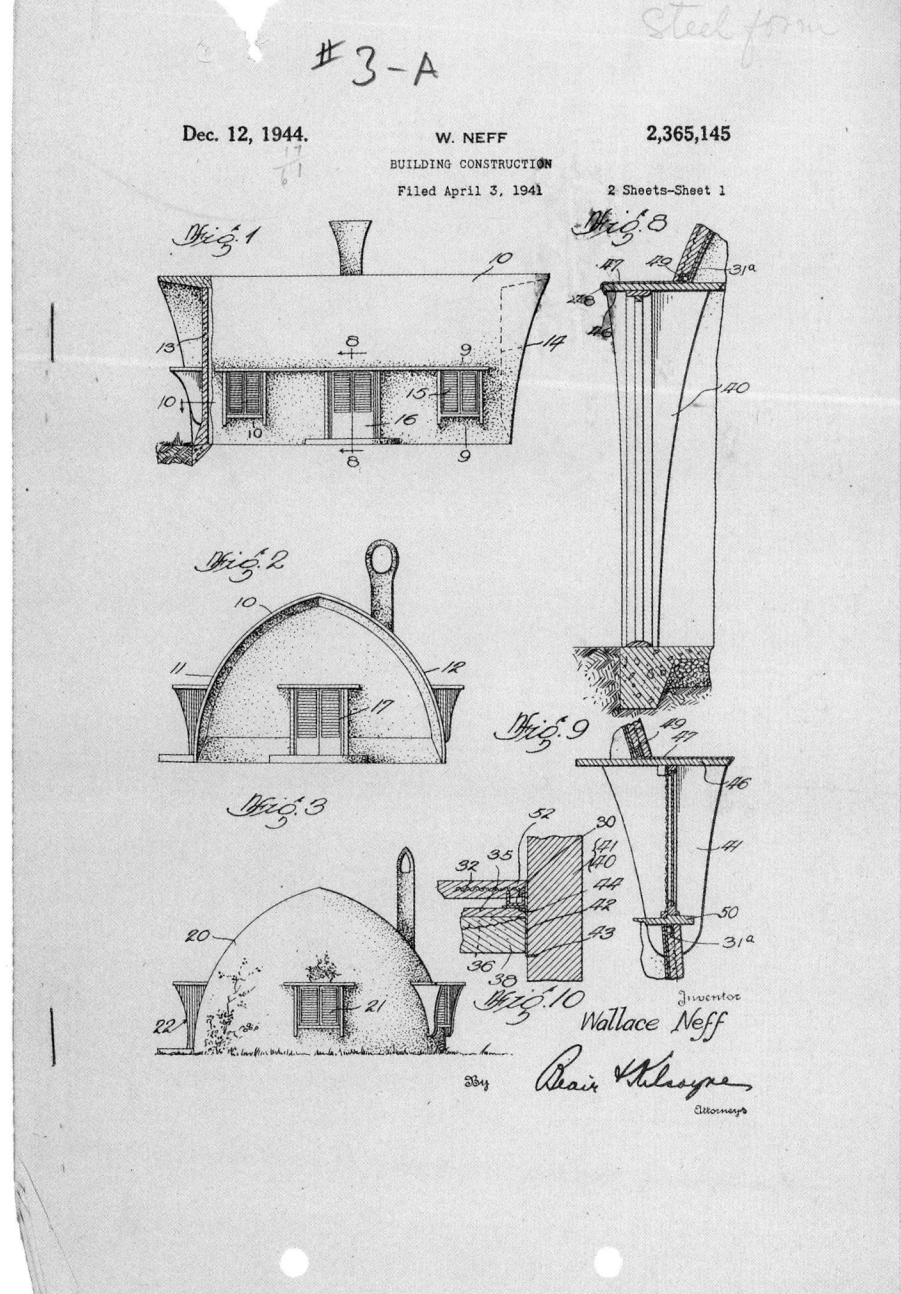

APPENDIX — B

Fig. 10: Neff's sketches of various Airform shapes

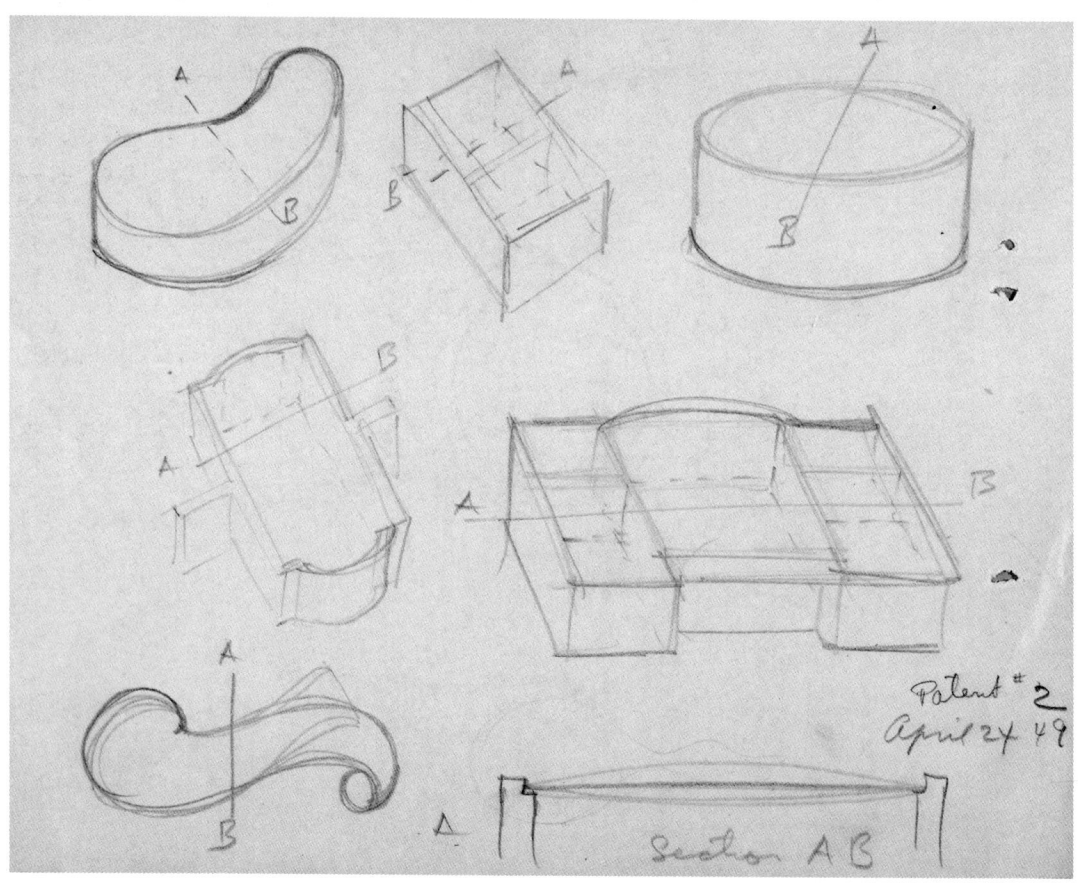

APPENDIX — B

Design for a House
Patent Number: D140060 Granted: January 16, 1945
A form with a single door and no windows.

Method and Apparatus for Constructing Shell-Form Structures
Patent Number: 2388701 / Granted: November 13, 1945
Details the open bottom Airform. Considered more efficient since it did not require as much of air pressure during construction. The balloon form could be made of nylon, cotton muslin instead of rubber-coated material or rubberized canvas—materials then in short supply during World War II. In 1942, the Cole of California, Inc., parachute plant in Vernon, California, was the first structure completed with this patent. Neff referred to the design as form "3."

Storage Tank and Method of Constructing
Patent Number: 2413243 / Granted: December 24, 1946
Specifies a new type of construction to replace concrete storage gasoline, oil, water, and other liquids, eliminating steel-reinforced concrete, to create a more cost-effective and efficient system of construction. Neff referred to the design as form "3-B" and it was associated with his underground bubble structures. [Fig. 10]

Improved Method of Erecting Shell-Form Concrete Structures
Patent Number: 2892239 / Granted: June 30, 1959
Details use of wire mesh to create reinforced, straight side walls that also offered control to shape of the form. Neff referred to the design as form "4."

APPENDIX — C

Selected Unbuilt Airform Projects, 1944–58

opposite
Fig. 11: An illustration of an unbuilt Airform for a public space

below
Fig. 12: Stylized floor plans for Unbuilt Airforms published in Mexico to promote to different uses

1944
▶ Five hundred bubble houses in the Channel Islands

1946
▶ Multiple plans for government housing in China
▶ Forty thousand bubble houses for the Philippine government

1947
▶ Various movie theaters in Mexico City, Mexico [**Fig. 12**]
▶ A lid for an industrial water tank for the Preload Corporation Houses in Greece for the Matrad Corporation and the Greek government
▶ A 2-million-gallon water tank in West Virginia and an 180-thousand-gallon reservoir for the Bridgeport Water Company in Norristown, Pennsylvania

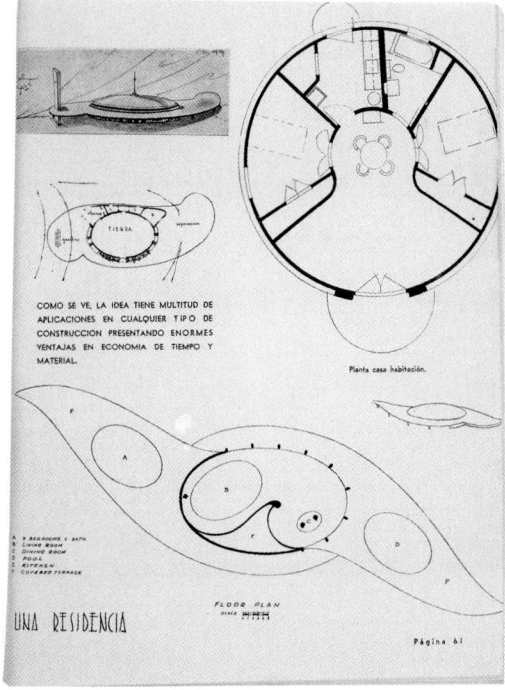

Fig. 13: Floor plan for a single-bubble Airform

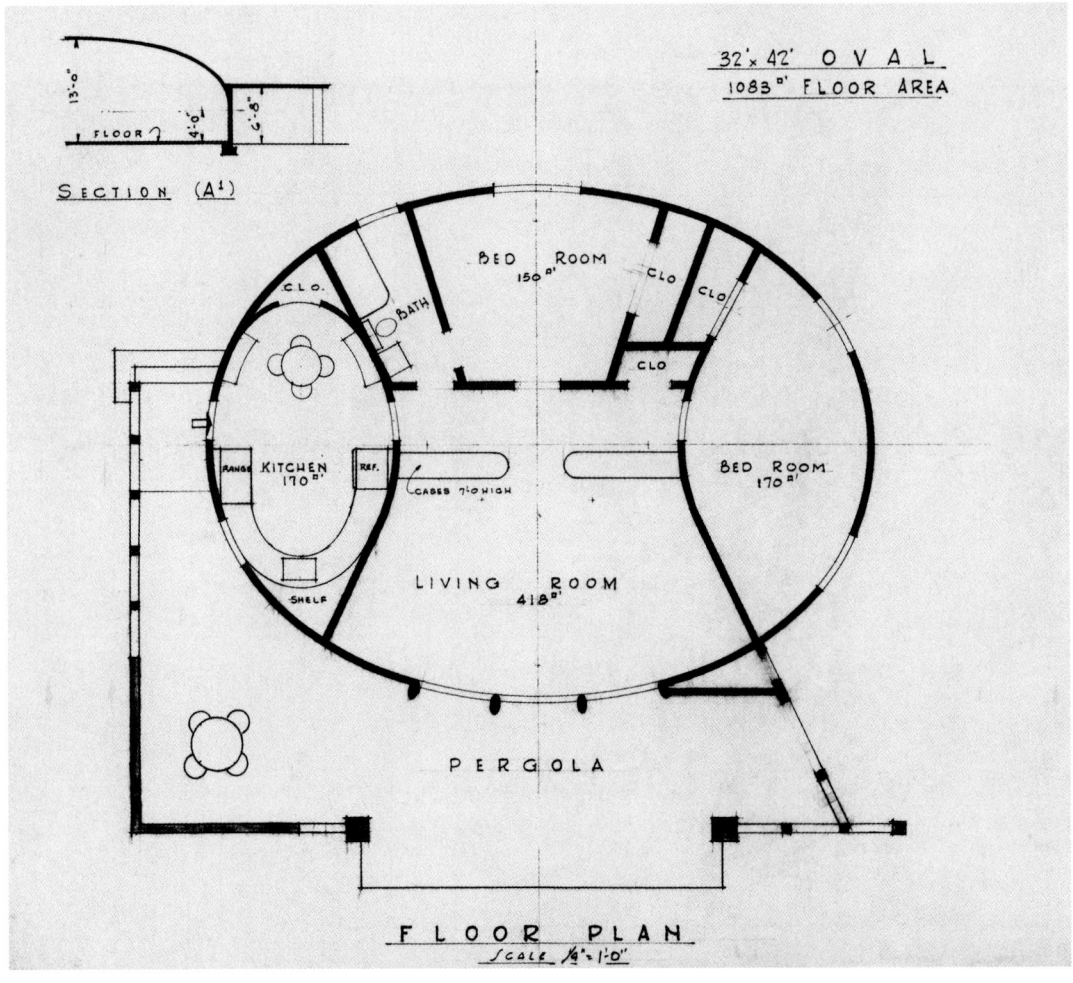

APPENDIX — C

1948

▸ Bubble houses for twenty thousand families for the U.S. government's atomic energy activities in Los Alamos, New Mexico [Fig. 13]
▸ A 1,200-seat theater for Twentieth Century Fox in Los Angeles, California

1949

▸ Various plans for residential and commercial use for the United States's Kadena Air Force base in Okinawa, Japan [Figs. 14 – 16]

1950s [Fig. 17]

▸ Dome covers for two- to five-million-gallon reservoirs for the city of Flint, Michigan
▸ Structures for atomic energy plants for the U.S. government
▸ A group of double-bubble houses in Richland, Washington
▸ A restaurant, "La Perla," in Mexico [Fig. 18]
▸ Three thousand bubble houses for the Preload Corporation and the Pacific Bridge Building Company of San Francisco to house factory workers in Tlalnepantla, Mexico [Fig. 19]

1951

▸ Various plans for La Mesa Housing in San Diego, with the LA Cement Gun Co.

1953

▸ One thousand bubble houses for public housing in Las Vegas, with Paul R. Williams
▸ A 350-room hotel in Las Vegas

1954

▸ Six thousand ammunition storage units in Bordeaux, France, for the French Air Force
▸ A motel on El Camino Real in San Mateo, California
▸ Large public housing projects by Harrison and Abramovitz and the IBEC Housing Corporation [Figs. 20 + 21]

below
Fig. 14: Unbuilt Airform military barracks

opposite top
Fig. 15: An illustration of an Airform military residence

opposite bottom
Fig. 16: Cut-away illustrations of an unbuilt Airform residence

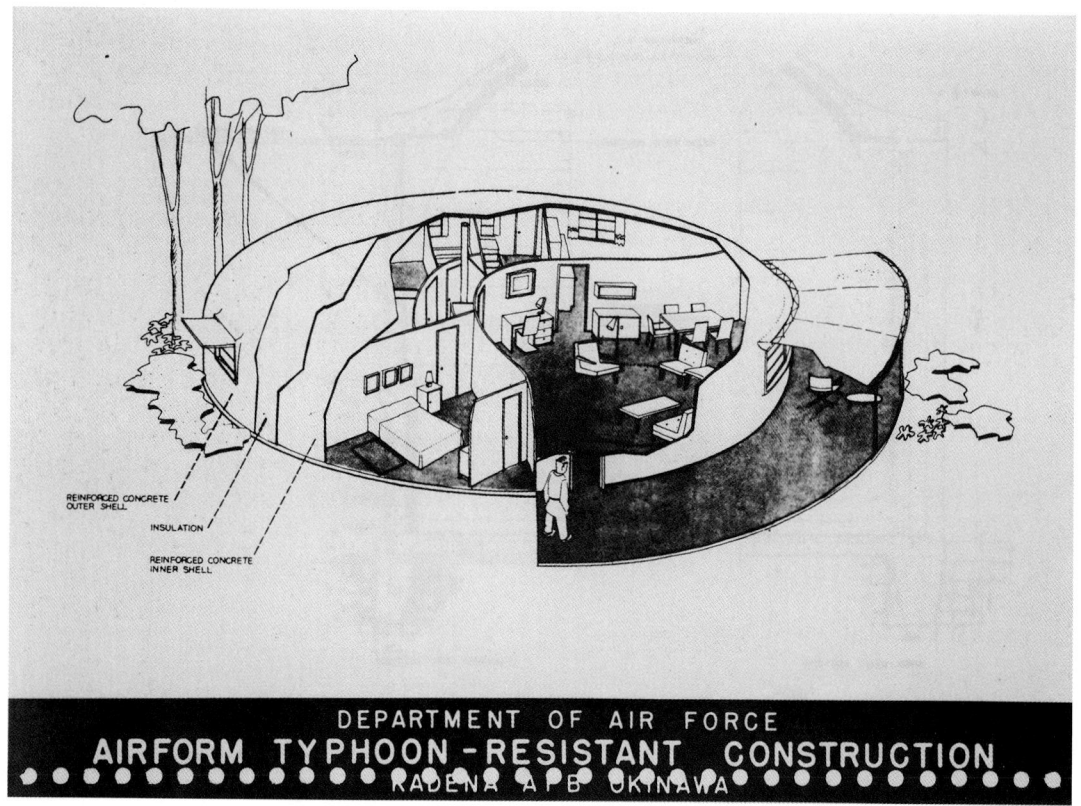

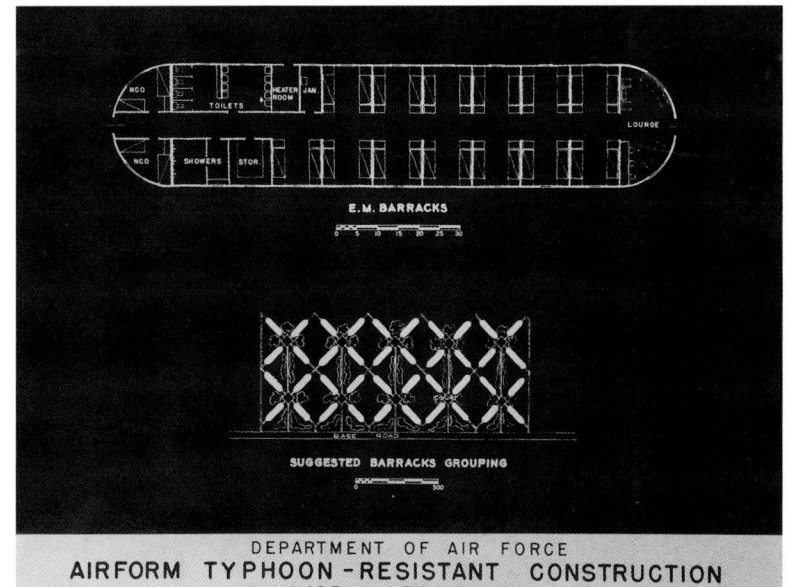

APPENDIX — C

below
Fig. 17: Rendering of an unbuilt Airform residence

opposite
Fig. 18: A hand-colored Airform design by Neff

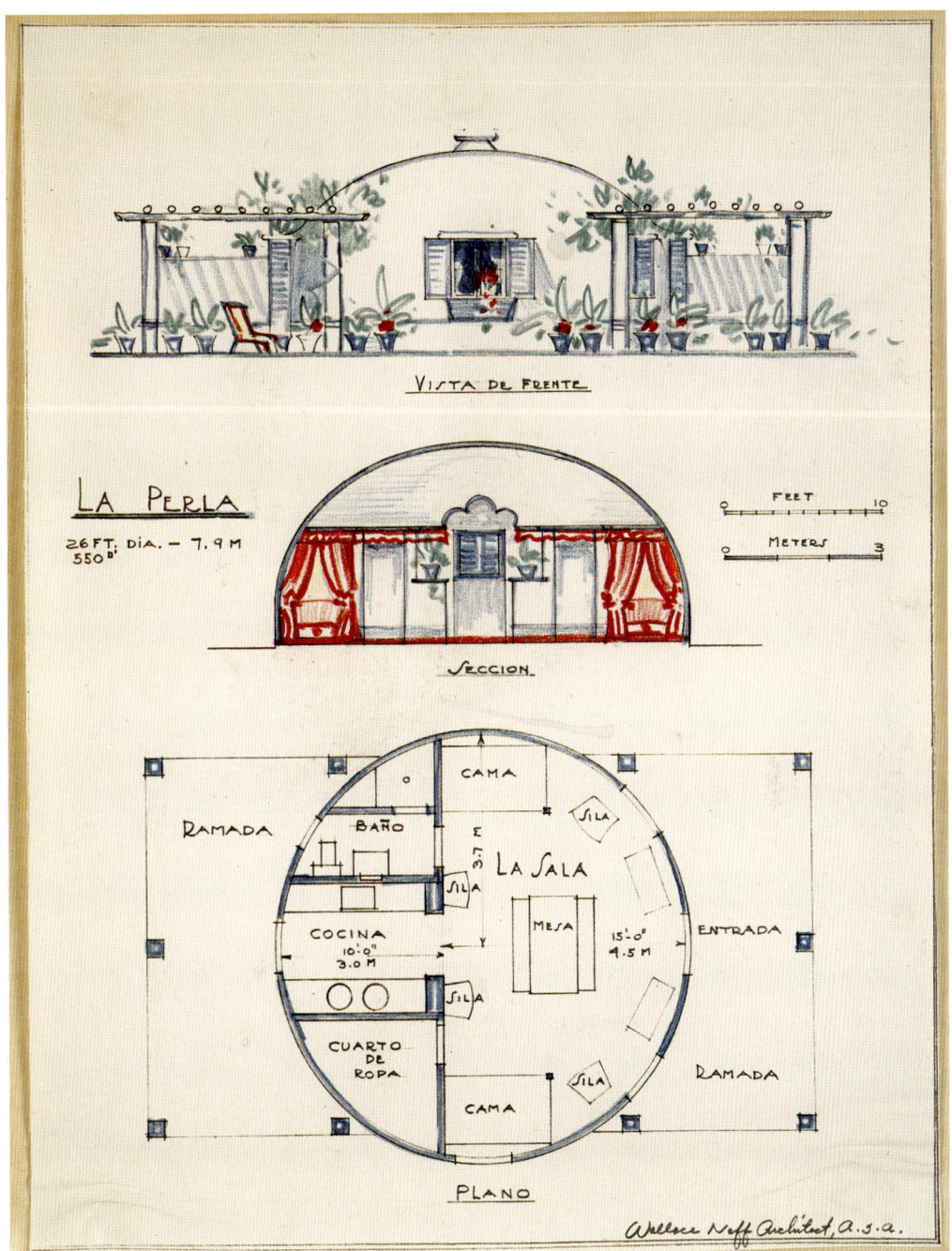

APPENDIX — C

right
Fig. 19: Data sheet for a three-bedroom Airform residence

opposite top
Fig. 20: Line-item cost estimates for high-volume Airform production

opposite bottom
Fig. 21: Additional accounting of cost estimates for high-volume Airform production

DATA SHEET

Airform Three Bedroom Farm Worker Home:

Area:

Sleeping	334 square feet
Living and Dining Areas	315 square feet
Kitchen	63 square feet
Bath	86 square feet
Hall	9 square feet
Storage*	16 square feet
Closet	22 square feet
Living Area	845 square feet
Wall Space and Utilities	25 square feet
TOTAL ROOFED AREA	870 square feet

*Additional storage space above closets and bedroom entrances – 55 square feet.

Features:

Shower dressingroom with outside entrance
Sliding closet doors
Storage cabinets above sink and stove
Kitchen shelves, enclosed kitchen sink
Stove, hot water heater, forced heat furnace – gas
Sound and heat insulation – 1/2 inch vermiculite under dome
Height – 11 feet 6 inches – allowing three layer bunks in bedrooms
Thin coat plaster or gunnited concrete partitions – absolutely durable, unaffected by fire, earthquake, termites, or vandalism.

Estimated Cost:

Approximately $6.50 per square foot in lots of 100 units in most areas of California.

APPENDIX — C

COST ESTIMATE

2 BEDROOM UNIT - 29' DIAMETER

COMPLETE WITH UTILITIES, ETC.

BASED ON CONSTRUCTION OF 250 OR MORE

Item	Unit	Labor	Material	Quantity	Material Cost	Labor Hours Skilled	Labor Hours Common	Labor Hours Total	Labor Cost	Total Cost
*Excavation	C.Y.	4.00		14			35	35	56	56
*Concrete Placing	C.Y.	4.00		12-3/4		7	23	30	51	51
*Concrete Mixing	C.Y.	3.00		12-3/4		4	19	23	38	38
*Cement	Bbls.	3.75		69	259					259
*Sand-Aggregate	C.Y.		2.50	40	100					100
*6 x 6 Mesh- 10 Ga.	S.F.	0.02	0.03	800	24	7		7	16	40
*2 x 2 Mesh- 18 Ga.	S.F.	0.02	0.06	2000	120	19		19	40	160
*Chicken Wire	S.F.	0.02	0.02	2000	40	19		19	40	80
*Reinforcing	Lbs.	0.055	0.055	440	24	11		11	24	48
*Metal Lath	S.Y.	0.50	1.00	97	97	21		21	49	146
*Shell & Partitions	C.Y.	11.50	2.50	19-1/2	49	58	58	116	224	273
Interior Doors	Ea.	4.00	27.00	9	243	18		18	42	285
*Exterior Doors	Ea.	4.50	32.00	2	64	4		4	9	73
*Metal Windows	S.F.	0.59	1.41	(8) 96	135	25		25	57	192
*Metal Screens	Ea.			8	48	15		15	21	69
Cement Floor Fin.	S.F.	0.06	0.04	660	27	19		19	40	67
*Membrane-Sisalkraft	S.F.	0.01	0.03	730	22	4		4	7	29
*Cement Wash	S.F.	0.06	0.10	1170	117	30		30	70	187
Asphalt Tile	S.F.	0.10	0.14	660	93	28		28	66	159
*Interior Paint	S.F.	0.05	0.03	3070	92	65		65	154	246
Window Shades	Ea.	0.22	4.00	8	32	1		1	2	34
*Runner Channel	L.F.	0.04	0.05	114	6	2		2	5	11
*Insulation	S.F.	0.04	0.07	1170	82	20		20	47	129
Oil Tank - 200 Gal.	Ea.		40		40	20		20	40	80
Oil Circul. Htr. 62000 BTU - Gravity	Ea.		140.00		140	10		10	20	160

COST ESTIMATE

2 BEDROOM UNIT - 29' DIAMETER

COMPLETE WITH UTILITIES, ETC.

BASED ON CONSTRUCTION OF 250 OR MORE

Item	Unit	Labor	Material	Quantity	Material Cost	Labor Hours Skilled	Labor Hours Common	Labor Hours Total	Labor Cost	Total Cost
Elec. Water Htr.	Ea.	13.00	90.00	1	90	3	3	6	13	103
Lavatory	Ea.	20.00	60.00	1	60	8		8	20	80
Water Closet	Ea.	20.00	113.00	1	113	8		8	20	133
Comb. Tub & Shower	Ea.	20.00	80.00	1	80	8		8	20	100
Kitchen Sink	Ea.	20.00	81.00	1	81	8		8	20	101
Laundry Trays	Ea.	20.00	45.00	1	45	8		8	20	65
Brkts, Shade, Curtain	Ea.	1.00	3.00	8	24	2		2	5	40
Medicine Chest	Ea.	5.00	35.00	1	35	2		2	5	40
Tissue Holder	Ea.	2.00	4.00	1	4	1		1	2	6
Towel Bars	Ea.	2.50	4.00	2	8	2		2	5	13
Shower Rod	Ea.	2.00	3.00	1	3	1		1	2	5
Mailbox	Ea.	1.00	2.00	1	2)			1	1	3
House Numbers	Ea.	1.00	1.00	1	1)	1		1)	1	2
Kitchen Cabinets	Ea.	37.00	286.00	1	286				37	323
Garbage Receptor	Ea.	5.00	15.00	1	15	2		2	5	20
Toilet Seat	Ea.		10.00	1	10					10
Soap & Grab Bar	Ea.		5.00	1	5)			1)	1	6
Soap Dish	Ea.		2.00	1	2)	1		1)	1	3
Coat Hooks	Ea.		3.00	2	6	1		1	2	8
Closet Poles	L.F.				6	1		1	2	8
Closet Shelves	S.F.		1.00	24	24	3		3	7	31
Builders Hdwe	Lump				50	7		7	16	66
Canopy Posts					103	13		13	31	134
Canopy Cover					40	6		6	14	54
Rough Plumbing,) Pipe Fitting)	Lump				67	16		16	40	107
*Flashing Roof Vent					16	7		7	15	31
Electrical					150	20		20	50	200
*Forms & Equip.					101	40		40	85	186
*Screen Doors	Ea.			2	14	2		2	4	18
					3281			716	1561	4860

* Items indicating 2 Br Basic Unit - 22'8" dia. only

APPENDIX — C

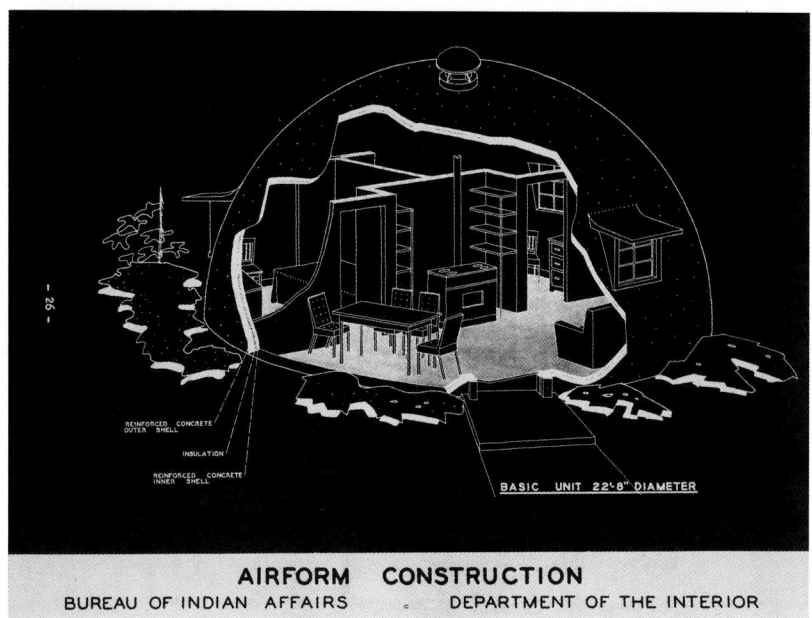

opposite top
Fig. 22: An unbuilt Airform for Native American housing

opposite bottom
Fig. 23: Cut-away illustrations of an unbuilt Airform residence

1955
▸ Thirty-five thousand bubble houses for American-Indian reservations [**Figs. 22 + 23**]

1956
▸ Airform containment structure for a nuclear reactor facility for Holyoke, Massachusetts
▸ Storage builings for the Union Tank Car Co., Chicago

1957
▸ Grain storage bins for the Italian Federation of Farmers Cooperative
▸ Bomb shelters for the U.S. Civil Defense Authority

1958
▸ Vegetable and grain storage bins for the General Foods Company in Eslou, Sweden

Afterword

During Neff's lifetime approximately 400,000 bubble houses were speculated for construction around the world. A fraction of these reached various planning stages that in the end lead to the completion of fewer than 2,500 houses and buildings.

The unusual design of the Airform structures made widespread acceptance difficult, particularly in the United States. Living in a bubble house required a change in lifestyle and raised concerns about resale value. Since there were no internal walls to define interior spaces, the space was largely left open and the floor plan may have been too unfamiliar for home buyers. The shape, room arrangement, placement of furniture, and other decorative concerns also challenged the Airform's aesthetics and for some, functional options. The limit of a single story may have also added to questions of livability since the form was not flexible and could not accommodate additions or expansions. Ultimately it was difficult for bubble houses to compete with traditional aesthetics and conventional housing of the era. Perhaps a social and cultural stigma also affected consumer perceptions.

From a business and government standpoint, the various shifts in administrations, with the associated changes with budgets and policies, contributed to a lack of commitment from Airform's international clients. Even when the best intentions of providing bubble houses served both public and private sectors, large-scale adoption was never forthcoming. Typically the completion of an Airform structure would gain the attention of local commercial interests and various government agencies, as with Constructora Airform de Cuba S.A, a company specifically formed to market and build Airforms. Two bubble houses were built in Havana but the company was unable to secure further support. In another situation, Construtora Airform S.L., a Spanish licensee, planned to build Airform storage containers for wine and olive oil, similar to those built in Portugal, but none were built even though the project in Portugal was a proven success and set a precedence for the wine industry.

When bubble houses were considered for emergency housing—disaster relief—a lack of qualified builders and contractors prevented plans from moving forward although the Airform's relatively simple construction, low cost, and use of available materials was consistently recognized.

Today, seventy years after Neff built the first bubble houses, a widely adopted housing solution remains elusive for today's architects and builders, however Neff effectively nailed it with the Airform.

Acknowledgments

This book was made possible with a generous research grant from the Graham Foundation for Advanced Studies in the Fine Arts. It is an honor for me to be a recipient of the award. Writing about Wallace Neff's bubble houses could not have happened without the Graham Foundation's support.

The most significant aspects of my research were conducted at the Huntington Library in San Marino, California, the repository of the Wallace Neff Collection, which formed the foundation of this book. Erin Chase, Curatorial Assistant, was always attentive and fully understood my research. Many additional images appear in the book because of her.

Scott Tennent, who might consider his involvement minor, was instrumental in guiding me through the development and manuscript phase. Thank you to Jennifer Thompson for seeing the potential of my work and for her patience with me. I worked closely with Dan Simon whose editorial direction was straightforward and helped me focus and refine the scope of writing for a more distilled manuscript. I understand the process more, thanks to him.

Thanks to Neil Bethke at the Loyola Marymount University Archives and Special Collections in Los Angeles; James Robert Allen II at Arizona State University–Architecture and Environmental Design Library; Sara Homan, Linda Lamm, and Rosemary Lang-Fiebig of the Litchfield Park Historical Society; and Craig Holbert at the University of Akron, Ohio.

I offer special acknowledgment to Stefanos Polyzoides, whose articulate nature offered insights about Neff I had not previously considered. I am indebted to the time and effort he gave me. Thanks also to photographers Leslie Williamson, Candace Feit, and Ron Rosenzweig.

I met many people in the course of writing this book that I would not otherwise know. For that, I am grateful to those who added interest and expanded my world: Virginia Tanzmann, Richard Tanzmann, Mary Mayhew, Kathy Miles, Arif Hasan, William Turner, Victor Miguel, Adrienne Wong, Onnis Luque Rodriguez, John Kelly, Ron Russell, Greg Packham, Frank P. Tighe Jr., and Michael C. Tighe.

My most personal thanks to: John and Marilyn Neuhart, whom I continue to learn from; Wallace L. Neff and Richard Pettigrew; Ann Gray for her early and helpful support; Yolanda Lopez for her translations of French and Spanish texts; Robin Purvis; Aaron Sosnick, friend and unwitting patron; Becky Fischbach; and SLPTL.

With gratitude I offer this book to Steve and Sari Roden.

Notes

Introduction

1. "Airform Construction Wallace Neff Architect," undated document, Airform Construction folder, box 2, Wallace Neff Collection, Huntington Library, San Marino, CA.
2. Ibid.
3. *The Book of Knowledge* (The Grolier Society: New York, 1943), s.v. "housing."
4. "Airform International Construction Corporation," undated marketing brochure, Airform Construction folder, box 2, Wallace Neff Collection, Huntington Library, San Marino, CA.
5. Robert L. Davison, *The Engineered Dwelling* (Raritan, NJ: The John B. Pierce Foundation, 1943), 6.
6. *The Book of Knowledge*, s.v. "housing."
7. "General Outline of Airform Construction."
8. "Airform International Construction Corporation," 4.
9. Ibid.
10. Ibid.
11. "General Outline of Airform Construction," undated document, Airform Construction folder, box 2, Wallace Neff Collection, Huntington Library, San Marino, CA.
12. "Plastic Bubble Buildings," March 1944, Airform Construction folder, box 2, Wallace Neff Collection, Huntington Library, San Marino, CA.
13. "Balloon Houses Designed for Defense Workers Bloom Under Virginia Tree," *Life* (December 1, 1941): 34–35.
14. *Wallace Neff 1895–1982: The Romance of Regional Architecture* (Huntington Library, San Marino, CA: 1998), 83.
15. "General Information Airform Construction," undated document, Airform Construction folder, box 2, Wallace Neff Collection, Huntington Library, San Marino, CA.
16. "Inflated Balloon Provides for Novel Defense House," *Architect and Engineer* (January 1942): 22–23.
17. *The Book of Knowledge*, s.v. "housing."
18. "The Pneumatic Form by Wallace Neff," September 1958, Airform Construction folder, box 2, Wallace Neff Collection, Huntington Library, San Marino, CA.
19. "A New Technique in Home Building." *National Real Estate Journal* (January 1942): 32–33.
20. "Airform International Construction Corporation."
21. Letter to Wallace Neff from J. A. Wahler, September 25, 1954, Airform Correspondence, box 2, Wallace Neff Collection, Huntington Library, San Marino, CA.
22. Ibid.
23. Licensee agreement between Wallace Neff and Adolph Waterval, October 3, 1955, Airform Construction folder, box 2, Wallace Neff Collection, Huntington Library, San Marino, CA.
24. Letter to Wallace Neff from Jose Lemos, August 18, 1956, Airform Correspondence, box 2, Wallace Neff Collection,

Huntington Library, San Marino, CA.
25. Ibid.
26. Licensee agreement between Wallace Neff and Robert N. Kuhn, January 29, 1963, box 3, Wallace Neff Collection, Huntington Library, San Marino, CA.
27. Untitled and undated document, Patents and Contracts folder, box 1, Wallace Neff Collection, Huntington Library, San Marino, CA.

Falls Church, Virginia
1. Patent notice, *New York Times*, January 25, 1942, D8.
2. "The Home built upon a BALLOON," [Goodyear Mechanical Goods Advertisement] *Newsweek* (January 1942): 34–35; and *Time* (January 5, 1942): 36–37.
3. "Ballyhooed Balloon," *Architecture Forum* (December 1941): 421.
4. "A New Technique in Home Building," *National Real Estate Journal*, (January 1942): 32–33.
5. "Ballyhooed Balloon," *Architecture Forum*.
6. Wallace Neff handwritten daily notes folder 1941–1951 [1941] Box 11. Huntington Library Wallace Neff Collection.
7. Letter dated April 4, 1986 from Douglas Fairbanks Jr. to Wally [Wallace L.] Neff. Wally [Wallace L.] Neff Collection.
8. Wallace Neff handwritten daily notes folder 1941–1951 [1941] Box 11. Huntington Library Wallace Neff Collection.
9. "Inflated Balloon Provides for Novel Defense House," *Architect and Engineer* (January 1942): 22–23.
10. "Balloon Houses," *Catholic Digest* (January 1942): 80.
11. Douglas Haskell, "Bubble House Afflatus," *The Nation* (February 28, 1942), 264–66.
12. Clark M. Bacon, "For the Bride of 1952," *Hollands—Magazine of the South* (April 1942): 6.
13. "Bubble Houses," *The Architects' Journal* (January 22, 1942): 72–74.
14. Hubert G. Ripley, "Back to pithecanthropus erectus: some notes on the igloo house," *Michigan Society of Architects* (January 6, 1941): 1.
15. Interview with William Turner, April 19, 2008.

Litchfield Park, Arizona
1. "Grain Bins for South West Cotton Company, Arizona, U.S.A." *Architect & Building News* (April 23, 1943): 60.
2. Ibid.
3. "Airform House for a Desert Colony," *Architectural Record* (July 1944): 81–83.
4. *Liberty*, September 29, 1945, 30.
5. "Airform House for a Desert Colony," 81–83.
6. Interview with Linda Lamm, July 22, 2008, with the author.
7. Wigwam Resort Brochure, undated, Litchfield Park Historical Society.

Loyola University, 1944
1. University Archives, Special Collections, Campus Files, President Records (Series 8), Loyola Marymount University.
2. Ibid.

Pacific Linen Supply Co.
1. Letter from Wallace Neff to "Mr. Robeson" [client], September 28, 1943, box 4, Wallace Neff Collection, Huntington Library, San Marino, CA.
2. Wallace Neff handwritten daily notes folder 1941–1951 [1943] Box 11. Huntington Library Wallace Neff Collection.
3. "Inflated—bag forms let it down," *Engineering News Record* 131 (December 2, 1943): 55.
4. "Concrete Dome," letter to editor by Wallace Neff, *Engineering News Record* 133 (August 10, 1944): 77.

Andrew Neff House, California
1. Note by Andrew Neff, undated, box 9, Wallace Neff Collection, Huntington Library, San Marino, CA.
2. Interview with Wallace Neff by Alson Clark and Jae Carmichael, Pasadena Historical Society and Friends of the Pasadena Public Library for the Pasadena Oral History Project, 1977.
3. "Unusal Dome–icile," *Pasadena Star News (*June 7, 1962): 25.
4. Ibid.

South Pasadena, California
1. Letter from William S. Stokes to Wallace Neff, May 14, 1946, box 3, Wallace Neff Collection, Huntington Library, San Marino, CA.
2. Interviews with Virginia Tanzmann and Richard Tanzmann, August 12 and August 19, 2008.

Hobe Sound, Florida
1. "Houses Sprayed from Gun," *Christian Science Monitor* (January 15, 1954): 12.
2. "Big glassed openings give balloon houses a new look," *House & Home* (March 1954): 153.
3. "Houses Sprayed from Gun," *Christian Science Monitor*.
4. "The Airform House," *Vogue* (January 1954): 127.
5. "Big glassed openings give balloon houses a new look," *House & Home*.

Latin America
1. "General Information Airform Construction," undated document, Airform Construction folder, box 2, Wallace Neff Collection, Huntington Library, San Marino, CA.
2. Licensee agreement between Wallace Neff and Joseph. A. Wahler, September 17, 1954, box 2, Wallace Neff Collection, Huntington Library, San Marino, CA.

Europe
1. "General Outline of Airform Construction," undated, Airform Construction folder, box 2, Wallace Neff Collection, Huntington Library, San Marino, CA.
2. Ibid.
3. "General Outline of Airform Construction."
4. Ibid.
5. Letter to Wallace Neff from J. A. Wahler, September 25, 1954, Airform Correspondence, box 2, Wallace Neff Collection, Huntington Library, San Marino, CA.

Africa
1. "Airform Construction General Information," undated marketing brochure, folder 3, box 5, Wallace Neff Collection, Huntington Library, San Marino, CA.
2. Interview with Wallace Neff by Alson Clark and Jae Carmichael, Pasadena Historical Society and Friends of the Pasadena Public Library for the Pasadena Oral History Project, 1977.
3. "Airform Construction General Information," undated marketing brochure, folder 3, box 5, Wallace Neff Collection, Huntington Library, San Marino, CA.
4. "Airform International Construction Corporation," undated marketing brochure, Airform Construction folder, box 2, Wallace Neff Collection, Huntington Library, San Marino, CA.
5. "Advantages of Airform Construction," proposal to the Department of the Air Force, undated, box 5, Wallace Neff Collection, Huntington Library, San Marino, CA.
6. An undated document outlining Wallace Neff's plans for different Airforms, Airform Construction folder, box 5, Wallace Neff Collection, Huntington Library, San Marino, CA.

Appendix A
1. Interviews with Mary Mayhew and Kathy Miles, March 18 and 21, 2008, with the author.

Appendix B
1. Wallace Neff. Building construction. U.S. Patent 2,365,145, filed April 3, 1941, and issued December 12, 1944.

Selected Bibliography

Several previously published bibliographical references do not appear in the following list. Their exclusion is not an omission or oversight.

1941

Architecture Forum, December 1941, p. 421
Architectural Record, December 1941, p. 22, 108
Life, December 1, 1941, p. 34–35
Los Angeles Times, November 23, 1941, p. 5
Architectural Forum, February 1941, p. 87–90.
Michigan Society of Architects, January 6, 1941, p. 1

1942

Portland Cement Association, 1942, p. CP50
Architect and Engineer, January 1942, p. 20–23
Catholic Digest, January 1942, p. 80
National Real Estate Journal, January 1942, p. 32–33
Newsweek, January 1942, p. 34–35
Western Construction News, January 1942, p. 12
Time, January 5, 1942, p. 36–37
Illustrated London News, January 10, 1942, p. 56
The Architects' Journal, January 22, 1942, p. 73–74
New York Times (patent notice), January 25, 1942, p. D8
The United States News, February 13, 1942, p. 24–25
The Nation, February 28, 1942, p. 264–266
Concrete Builder, Spring 1942, p. 8–9
Fortune, March 1942, p. 155
Popular Mechanics, March 1942, p. 155
Hollands—Magazine of the South, April 1942, p. 6
Nuestra Arquitectura, May 1942, p. 174–177
Los Angeles Times, June 29, 1942, p. 5
Arizona Farmer, July 18, 1942, p. 3
Christian Science Monitor, July 31, 1942, p. 10
Christian Science Monitor, September 2, 1942, p. 17

1943

Architectural Forum, February 1943, p. 76–78
Architect & Building News, April 23, 1943, p. 60
Engineering News Record, December 2, 1943, p. 55
Davison, Robert L. *The Engineered Dwelling*. Raritan, NJ: The John B. Pierce Foundation, 1943.
The Book of Knowledge. New York, NY: The Grolier Society, 1943

1944

The Indiana Framers Guide, January 1944, p. 10
Southwest Builder & Contractor, February 11, 1944, Cover
Engineering News Record, February 24, 1944, p. 97

Popular Mechanics, March 1944, p. 7
Forbes, April 1944, p. 26
Fortune, May 1944, p. 170
Popular Mechanics, May 1944, p. 46–48
Architectural Record, July 1944, p. 81–83
Country Gentleman, July 1944, p. 15
Southwest Builder & Contractor, July 28, 1944, Cover
London Pictorial, August 6, 1944
Engineering News Record, August 10, 1944, p. 77

1945
Popular Mechanics, April, 1945, p. 48
L'Architecture d'Aujourd'hui, July–August 1945, p. 4–12
Prefabrication, September 4, 1945, p. 38–40
Liberty, September 29, 1945, p. 30
Goodyear Triangle, November 20, 1945
Bruce, Alfred and Harold Sandbank. *A History of Prefabrication*. Raritan, NJ: The John B. Pierce Foundation, Housing Research, 1945.

1946
Picture Wise, May 1946, p. 22
See, September 1946, p. 24
Los Angeles Times, October 20, 1946, p. E5
Casson, Hugh. *Homes by the Million: An Account of the Housing Methods of the U. S. A. 1940–1945*. Harmondsworth, Middlesex, England: Penquin Books, 1946.
Forman, Robert. *Make it Yourself Architectural Models*. London & New York: The Studio, 1946.
Gloag, John, and Grey Wornum. *House Out of Factory*. London: George Allen & Unwin, 1946.

1947
Los Angeles Home Magazine, April 13, 1947, Cover
Revista de la Esquela National de Arquitectura, July 1947, p. 58–6
West Coast Real Estate & Business Opportunity Journal, July 1947, p. 10
Architectural Forum, July 1947, p. 10
A Rodovia, November 1947, p. 54

1948
Engineering News Record, April 22, 1948, p. 5

1949
Rubber Popular Blad Gewidaan Rubber, February 1949, p. 15
Engineering News Record, August 25, 1949, p. 34

1950
Journal of the American Institute of Architects, November 1950, p. 221
Christian Science Monitor, November 10, 1950, p. 14
Gloag, John. *Men and Buildings*, second edition, revised and illustrated. London: Chantry Publications Limited, 1950.

1951
Interiors, March 1951, p. 14
Engineering News Record, June 1951, p. 55

1952
Science Digest, September 1952, p. 20–22

1953
Reader's Digest, June 1953, p. 98–99
Time, June 22, 1953, p. 62
Prefabrication, November 1953, p. 22
Popular Science, December 1953, p. 133–135
Engineering News Record, December 24, 1953, p. 45

1954
Mademoiselle, January 1954, p. 91
Vogue, January 1954, p. 127
Christian Science Monitor, January 15, 1954 p. 12
Time, January 25, 1954, p. 104
Life, February 22, 1954, p. 75–76, 78
House & Home, March 1954, P. 153
American Industrial Exporter Special Report, 1954, p. 22
Architectural Record, May 1954, p. 314, 316
Progressive Architecture, June 1954, p. 116–119
Architectural Record, November 1954, p. 221, 223

1955
Arts and Architecture, January 1955, p. 12–15, 32–35
House & Home, January 1955, p. 134
Architectural Record, February 1955, p. 206–208

1956
Architectural Record, Mid-May 1956, p. 204–205
Domus, May 1956, p. 13–14

1957
Business Week, January 5, 1957, p. 206–208

1948
Business Week, August 23, 1958, p. 206–208

1959
Time, August 18, 1959, p. 24–26

1961
Junior Scholastic, January 25, 1961, p. 8

1963
Kaswell, Ernest R. *Wellington Sears Handbook of Industrial Textiles*. New York: Wellington Sears Company, Inc., 1963.

1964
United Nations Review, March 1964, cover
Neff, Wallace. *Architect, FAIA Architecture of Southern California Architecture of Southern California; a selection of photographs, plans, and scale details from the work of Wallace Neff*. Chicago: Rand McNally, 1964.

1976
Newlon, Howard Jr., ed. *A Selection of historic American papers on Concrete 1876–1926*. Detroit: American Concrete Institute, 1976.

1977
Clark, Alison and Jae Carmichael. Interview with Wallace Neff. Pasadena. Historical Society; Friends of the Pasadena Public Library; Pasadena Oral History Project. 1977.

1980
Bernhardt, Arthur D. *Building Tomorrow: The Mobile/Manufactured Housing Industry*. Cambridge, MA: MIT Press. 1980.

1982
Los Angeles Times, June 10, 1982, p. G9

1986
Neff, Wallace, Jr., and Alison Clark. *Neff, Architect of California's Golden Age*. Foreword by David Gebhard. Santa Barbara, California: Capra Press, 1986.

1987
Pasadena Star News, May 17, 1987, p. B–1

1989
Pasadena Star News, May 11, 1989, p. B–1
Wallace Neff, 1895–1982: The Romance of Regional Architecture. In connection with an exhibition presented by the Virginia Steele Scott Gallery at the Huntington Library from May 6 through September 4, 1989. Huntington Library, San Marino, California. 1989.

1995
World War II and the American Dream. Albrecht, Donald, ed. National Building Museum, Cambridge, MA: MIT Press. 1995.

1998
Architectural Design, September 1998, p. iii–iiv

2001
Rose, Joseph. *Folds blobs + boxes: Architecture in the Digital Era*. Pittsburgh, PA: The Heinz Architectural Center Books, 2001.

2005
Kanner, Diane. *Wallace Neff and The Grand Houses of the Golden State*. New York: Monacelli Press, 2005.

2006
Bruce, Gordon. *Eliot Noyes: A Pioneer of Design and Architecture in the Age of American Modernism*. New York: Phaidon Press Inc., 2006.

Image Credits

Huntington Library, Maynard Parker Collection: 2, 45 top,
   46–51, 53 bottom right, front cover image
Courtesy of the Huntington Library: 12, 14, 22, 25 right,
   26–30, 32–33, 35 bottom left, 36 left, 37–38,
   39 top right and middle left, 40, 53 top, 55, 57–59, 60 right,
   63, 69–71, 73 top, 75 top, 78 top, 79 top, 87 bottom, 89,
   91 top, 92, 95 top, 96 top right, 97, 98 bottom, 101 top, 102,
   103 bottom, 105–12, 115 bottom, 116 top, 124, 125 bottom,
   138–40, 142–44, 147–51, 153–56
Courtesy of Steve Roden: 16, 17 top, 19–20, 23–24, 36 right,
   39 middle right, 39 bottom left, 39 bottom right,
   43 top right and middle left, 44, 56, 88, 90, 91 bottom, 93,
   94, 95 bottom, 96 top left, 96 bottom, 98 top, 99–100,
   101 bottom, 114, 115 top, 123 top, 131
Courtesy of Wallace L. Neff: 10–11, 17 bottom, 25 left,
   35 top and bottom right, 41, 52, 62, 64, 72, 73 middle and
   bottom, 74, 75 bottom, 76–77, 78 bottom, 79 bottom,
   84–86, 113, 116–17, 123 bottom, 124 left, 124 top, 146,
   back cover images: top and bottom left, bottom right
Courtesy of the Litchfield Park Historical Society:
   43 middle right, 45 bottom
Courtesy of the Arizona State University Library: 43 bottom,
   87 top, 152
Courtesy of the Loyola Marymount University Library:
   53 bottom left
The *Los Angeles Times*: 60 left
Leslie Williamson: 61, 65–67
Robert Rosenzweig: 80–83, back cover image: top right
Private collection: 103 top
Candace Feit: 118–19
Victor Miguel: 121
Kathy Miles and Mary Mayhew: 128–32

FEB 2012 B W

DISCARD